HISTORY'S
WORST
INVENTIONS

and the People Who Made Them

HISTORY'S WORST INVENTIONS

and the People Who Made Them

Eric Chaline

NEW HOLLAND

Published in 2009 by New Holland Publishers (UK) Ltd
London • Cape Town • Sydney • Auckland
www.newhollandpublishers.com

Garfield House, 86–88 Edgware Road, London W2 2EA, United Kingdom
80 McKenzie Street, Cape Town 8001, South Africa
Unit 1, 66 Gibbes Street, Chatswood, NSW 2067, Australia
218 Lake Road, Northcote, Auckland, New Zealand

10 9 8 7 6 5 4 3 2 1

Copyright © Quid Publishing, 2008

Quid Publishing
Level 4,
Sheridan House,
114 Western Road,
Hove,
BN3 1DD
www.quidpublishing.com

A catalogue record for this book is available from the British Library

ISBN 978 1 84773 505 8

Reproduction by Universal Graphics
Printed and bound in Singapore

To Hilary, Pamela, Matthew and Deamion, good friends and helpmates.

CONTENTS

INTRODUCTION

It was long believed that our ability to make and use tools was what set us apart from other animals. Although it is now known that many animals have tool-making abilities, no other species has ever matched the breadth of our creativity and inventiveness in making things for our own benefit and betterment, as well as for our own harm and destruction.

In the context of this book, 'worst inventions' can have several meanings. The first is failure, be it heroic, tragic or ridiculous – though sometimes inventions have managed to be all three. Consider the tragic case of the airship *Graf Zeppelin* (see pp. 131–5), an amazing feat of technology that was fatally compromised because its inventors decided to cut costs by substituting non-flammable but expensive helium for the cheaper but highly explosive hydrogen. Failure could also have been because an invention was far ahead of its time, which was the case with Brunel's atmospheric railway (pp. 104–7) and perhaps the hybrid human–electric vehicle (pp. 244–6).

A second group of inventions have earned their place because of their unforeseen consequences: miracle materials, miracle cures and miracle drugs that have led to death and injury on an epic scale. Examples abound in the pages of this book: the 'miracle' materials, asbestos (pp. 16–20), radium (pp. 146–9) and polythene (pp. 141–5), and the 'wonder' drugs, heroin (pp. 136–40), LSD (pp. 183–7) and thalidomide (pp. 218–22). But unforeseen consequences can be far more insidious and long-term, as is the case with soda pop (pp. 70–4) and junk food (pp. 150–5), both major contributors to the obesity epidemic currently afflicting the developed world. Although there is often no one to blame for causing these unforeseen effects, they have often been made much worse by human stupidity, ignorance and greed. For example, the health risks of smoking (pp. 32–7) were understood in 1948, but it took decades to force tobacco companies to admit that their products were potentially fatal, and for governments to start acting to protect their citizens from harm. Similarly, the burning of fossil fuels (pp. 21–6) and the use of CFCs (pp. 193–7) continue despite their widely understood consequences for the welfare of the planet.

A third group is included because they serve no conceivable useful purpose other than flattering our vanity or taking advantage of our credulity. Examples of the former are fashion items such as the crinoline (pp. 99–103), the corset and the platform shoe (pp. 223–5) – items that condemned women to physical incapacity and often injury. Fraudulent therapies to improve health or appearance, increase male potency or extend life are good examples of the latter, as are the many get-rich-quick schemes based on complex financial instruments that we seem to have fallen for with tiresome regularity from the seventeenth-century mania for tulips (pp. 64–9) that nearly ruined the Dutch economy, to the modern-day credit crunch caused by the invention of the now infamous 'subprime' mortgage (pp. 247–51).

The fourth group of inventions that you will find here are those designed specifically to have the most unpleasant consequences for the individual or humanity as a whole. This category includes a fair amount of weaponry: from the switchblade that can kill one person at a time (pp. 75–8), to nuclear (pp. 203–8), chemical (pp. 54–9) or biological weaponry (pp. 79–84) designed to kill thousands or even millions. But purposefully destructive inventions are not limited to the weapons of war. Two examples immediately suggest themselves from the world of information technology: the twin scourges of computer viruses (pp. 226–30) and spam email (pp. 239–43) bring havoc to private lives and business alike.

Finally, the inclusion of certain inventions might surprise the reader. For example, karaoke (pp. 235–8) can only be described as an extremely successful phenomenon that has spread rapidly around the globe – that is, until you are subjected to a pitifully out-of-tune rendition of 'My Way'. Conversely, the reader may think that one of their pet worst inventions has been left out. I can only answer that there is no definitive list.

If this volume serves a purpose, it is to make the reader reflect on the brilliance, heroism, creativity, blind stupidity, pigheadedness, credulity, greed and self-destructiveness that characterise the human race in all its inventiveness.

FAILING

Never got off the drawing board

Didn't work in practice

Killed its inventor

A commercial failure

Unforeseen consequences

Was used for evil ends

A success born of failure

FLIGHTS OF FANCY: ICARUS AND THE DREAM OF HUMAN-POWERED FLIGHT

Main Culprits: Daedalus and later inventors

Motivation: Scientific inquiry; fame

Damage Done: Ridicule, injury and death

"'Let me warn you, Icarus, to take the middle way, in case the moisture weighs down your wings, if you fly too low, or if you go too high, the sun scorches them. Travel between the extremes. And I order you not to aim toward Bootes, the Herdsman, or Helice, the Great Bear, or toward the drawn sword of Orion: take the course I show you!" At the same time as he laid down the rules of flight, he fitted the newly created wings on the boy's shoulders.'

Ovid, 43 BCE–17 CE, *Metamorphoses*, trans. A.S. Kline

Humans (though in this case, 'men' might seem more appropriate – history records far fewer female pioneers of unaided flight) have always been fascinated by the ability of birds to fly. In ancient mythology, flight was a privilege of the gods and their servants, and humans imitated them at their peril. Flying animals, such as the winged horse Pegasus, and flying objects, such as magic carpets and broomsticks, abound in myth and legend, but humans flying under their own power are far more unusual. Looking at it from a philosophical perspective, human-powered flight could be seen as a metaphor for absolute freedom, and in the ancient Greek myth of Daedalus and Icarus – one of the earliest stories to deal with human-powered flight – freedom and flight are closely linked.

King Minos of Crete, so the myth goes, commissioned the architect and inventor Daedalus to build the palace of the labyrinth as a home for Minos's monstrous son, the bull-headed man-eating Minotaur. But once the palace was finished, Minos was loath to let Daedalus and his son Icarus leave, lest they reveal its secrets. Unable to escape the island by ship, Daedalus made wings of feathers held together with wax for himself and his son.

FLYING HIGH

In Ovid's version of the story (see quote), Daedalus warned Icarus of the dangers of flight, but the boy, enchanted by his new-found freedom, flew too close to the sun, with disastrous consequences. His wings melted, he fell into the sea and was drowned. However, in one important respect, the story of Icarus is untypical of the many attempts at human-powered flight that followed it; namely, the wings, though imperfect, worked, allowing Daedalus to fly all the way home to Sicily far across the sea from Crete.

Another legendary pioneer of human-powered flight was King Bladud of the Britons, who supposedly reigned around 860–840 BCE. The mythical founder of the English spa town of Bath, Bladud is said to have made a pair of feathered wings and launched himself from the tower of the Temple of Apollo in London, then known as New Troy. According to one account, he died colliding with a wall; in another, he broke every bone in his body upon hitting the ground.

Despite these two dire examples from legend, humanity has continued in its pursuit of unaided flight. In historical times, three approaches have won favour with inventors: the glider or parachute (which is strictly speaking not full human-powered flight, as the initial, and sometimes fatal, thrust is provided by gravity); the helicopter rotor principle; and the winged aircraft.

Two medieval pioneers of the glider/parachute approach were the Arab Abbas Ibn Firnas (810–887 CE), who lived in the Islamic kingdom of Cordoba in Spain, and the English monk Eilmer of Malmesbury, who lived around the start of the eleventh century.

Firnas survived two attempts at flight: In 852, he used a primitive form of parachute to break his fall when jumping from the minaret of the Great Mosque of Cordoba and escaped with minor injuries. Undeterred, in 875 he made a second attempt from a hill overlooking the city with an early form of hang glider, at what was then the extremely advanced age of 65. On this occasion, he crashed and badly injured his back, ending his flying career.

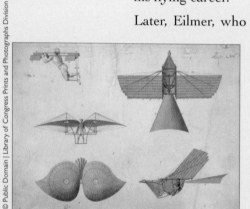

Later, Eilmer, who claimed to be inspired by the story of Daedalus, jumped from Malmesbury Abbey tower with wings strapped to his back. Eyewitnesses said that he travelled about 650 ft (200 m) before hitting the ground and breaking both his legs. His injuries left him lame and put an end to his aeronautic adventures.

A few centuries later, the Renaissance polymath Leonardo da Vinci (1452–1519) famously designed the first cone-shaped parachute; but, like many of his inventions, it was probably never built. A later

FLYING MACHINES
Technical illustrations showing Besnier's design for a flying machine (top left); two views of Jakob Degen's 'ornithopter' (centre and bottom left); and two views (top and bottom right) after designs for the construction of a flying car published in Thomas Walker's *The Art of Flying*.

model by the Croatian Faust Vrancic (1551–1617) was successfully tested in Venice by its inventor, who appears to have survived a jump from a tower unscathed. Then, in 1678, a French locksmith named Besnier created a simple flying machine (shown left) consisting of two wooden bars over the shoulders, with folding muslin shutters at either end. The distance that Besnier could cover by beating these 'wings' was limited, however, and it seems that the height of his achievements was to jump from an upper floor of his own house and sail clear over a nearby cottage.

Several such 'beating wing' machines continued to appear over the following years, for example, in the form of one 'ornithopter' – an aircraft with flapping wings that would imitate the flight of a bird – made by the Swiss clockmaker Jakob Degen in 1809, and the designs for a flying car that appeared in Thomas Walker's 1810 pamphlet *The Art of Flying* (both shown opposite).

A quite different approach was taken by the 'flying tailor', Franz Reichelt (c. 1860–1912), who jumped from the first viewing deck of the Eiffel Tower in Paris – a height of 196 ft (60 m) – wearing his own patent 'parachute overcoat'. A poignant video of the event is available on the Internet. Reichelt is shown standing on the balustrade of the tower's viewing deck before his final 'flight' ('headlong plummet' would be more accurate). The laconic French commentary reads: 'As if the unfortunate inventor sensed the horrible fate that awaited him, he hesitated for a long time before launching himself into the void.'

A much more successful hang-glider pioneer was the German Otto Lilienthal (1848–96), though, like many of his brave breed, he died at the tragically young age of 48. Unlike many of his predecessors, however, Lilienthal was an engineer who had a good practical grasp of aerodynamics. He studied the flight of birds and applied his discoveries to the design of the first successful hang gliders. A veteran of some 2,000 successful flights, he died when his glider was caught in turbulence, and he crashed from a height of 56 ft (17 m), fatally injuring himself.

TRUE FLIGHT

Although both the parachute and the hang glider have been perfected in the modern era, they give only the semblance of full human-powered flight. Without the external force provided by gravity, wind, a launching device or a powered craft of some kind, the parachute and glider remain stubbornly earthbound. A truly human-powered aircraft needs to be able to become airborne and remain aloft under its pilot's power.

Unsurprisingly, the great Leonardo did not stop at the parachute, but also applied himself to the problem of true human-powered flight. He came up with designs for the ornithopter and for the first helicopter, though his design was based on a helix and not a rotor blade as in modern machines. Neither of these was built at the time, and later reconstructions were incapable of flight because the materials he used

– canvas and wood – were far too heavy. Da Vinci, however, inspired many later inventors, who tried to apply these two approaches in their own creations. However, after the advent of powered aircraft at the beginning of the twentieth century, all of these attempts seemed doomed to remain eccentric, heroic failures.

The prospect of ignominious failure, however, did not discourage the twentieth century's pioneers of human-powered aviation. Many inventors have made use of the technology available at their times in their attempts to get airborne. The bicycle has proven an enduring and popular theme, as its efficient gearing system provides considerable mechanical advantage on the ground. Although these inventors were on the right track in using pedal power, the materials available to them were not suitable for building a viable flying machine. The conventional metal alloys, wood and canvas were far too heavy for the limited supply of human strength and stamina. So, Umberto Carnevali, Edward Frost, Sanderson Chirambo, Ernest Winter, Clifford Davis, Wally Smith, Bob Preston and many, many others, we salute your ingenuity and courage, even though they were accompanied by total and utter insanity.

If the secret of human-powered flight was ever going to be cracked, it was going to take some serious brainpower, a full understanding of aerodynamics, not to mention the latest in modern materials technology. All this would be expensive. So, to hurry things along, in 1959, the Royal Aeronautical Society of Great Britain instituted the Kremer Prize, to be awarded for the first human-powered flight.

The prize specified that to win, an aircraft would have to fly a figure-of-eight course of one mile (1.6 km), starting and finishing the flight at a height of no less than 10 ft (3 m) from the ground. The first prize was set at £50,000 (US $75,000–100,000), which represented a considerable sum in the late 1950s. However, even with this large financial inducement, it was 18 years before the first prize was won.

THE FIRST KREMER PRIZE TO BE AWARDED WAS WON BY *GOSSAMER CONDOR*, PILOTED BY BRYAN ALLEN IN 1977.

The first of three Kremer prizes to be awarded to date was won by Paul MacCready and Peter Lissaman's *Gossamer Condor*, piloted by cyclist and hang-glider pilot Bryan Allen, who successfully flew the specified course on August 23, 1977. The *Condor* had a wingspan of 96 ft (29 m). Despite its size, the craft, built of lightweight plastics, weighed a mere 70 lb (32 kg). The pilot sat in a gondola suspended from the wing, and powered the large rear propeller turned by a pedal-and-gear arrangement.

The same team won the second Kremer prize of £100,000 (US $175,000–200,000) when Allen flew the *Gossamer Albatross* the 22 miles (35.4 km) across the English Channel on June 12, 1979. The *Albatross* was roughly the same size and weight as the *Condor*, except that the gondola was enclosed in polythene film. The aircraft completed the crossing to France in less than three hours, clocking up an average speed of 18 mph (29 km/h), but skimming the waves at a height of just 5 ft (1.5 m). The final Kremer prize was awarded for a speed trial to a team from Massachusetts Institute of Technology (MIT), whose craft, the *MIT Monarch B*, succeeded in flying a course of 0.9 miles (1.5 km) in under three minutes. There are a further three Kremer prizes yet to be claimed, including one for flying a marathon course (26 miles/42 km) in under an hour. The longest human-powered flight on record was achieved on April 23, 1988, appropriately enough in the Aegean, when the *MIT Daedalus* flew the 74 miles (119 km) from Crete to the island of Santorini.

The Sikorsky Prize was established in 1980 to promote the first human-powered helicopter flight. In contrast to the fixed-wing aircraft successes, the current world record for a human-powered helicopter, the Japanese-built *Yuri-1*, stands at a height of 7.8 in (20 cm) for a duration of 19.46 seconds.

HISTORY'S FINAL VERDICT

So, humans have succeeded in flying under their own power, at least within the criteria set by the Royal Aeronautical Society. It has proven possible for a human, with the strength and stamina of an Olympic athlete and equipped with a very special type of aircraft, to get airborne and, given the right climatic conditions and a fair wind, to travel over great distances.

However, the technology, although demonstrating humanity's incredible ingenuity and creativity, has no practical applications. You will never see Joe Average hauling a pedal-powered aircraft from his garage for his daily commute to the office – even if the price of oil rockets sky-high. Think of the congestion and parking problems alone!

FAILING

Never got off the drawing board

Didn't work in practice

Killed its inventor

A commercial failure

Unforeseen consequences

Was used for evil ends

A success born of failure

THE MIRACLE MINERAL: CHARLEMAGNE'S ASBESTOS TABLECLOTH

Main Culprits: Asbestos mining and manufacturing companies and national governments

Motivation: Greed

Damage Done: Slow and painful deaths for millions of workers

'Asbestos looks like alum and is completely fireproof; it also resists all magic potions, especially those concocted by the Magi.'

Pliny the Elder (23–79 CE), *Naturalis Historia* (*Natural History*) trans. J.F. Healy

The ancient Greeks and Romans knew about the fire-resistant properties of asbestos (from the ancient Greek for 'inextinguishable'), which led them to call it the 'miracle mineral'. They regarded it as a quasi-magical substance that could ward off black magic, but they were also aware of its dangers.

One of the earliest uses for the fibrous mineral was in the manufacture of cloth, and the Roman naturalist Pliny the Elder (23–79 CE) was among the first to note that asbestos weavers suffered from illnesses of the lungs. Later the Holy Roman Emperor Charlemagne (747–814 CE) had a tablecloth made of asbestos, and his favourite after-dinner party trick was to throw the cloth onto the fire to clean it – in those days, you had to make your own entertainment. Other preindustrial uses for the miracle material were everlasting wicks for funerary lamps, and cremation shrouds, which allowed the body to be burned without its remains getting mixed up with the ashes from the pyre.

Asbestos remained something of an oddity until the nineteenth century, when it found many uses in both industrial and consumer products. Although the health risks of working with asbestos were understood in the late nineteenth century, it took a further hundred years for asbestos products to be controlled in much of the developed world. However, such is the influence of big business in the United States that asbestos is still found in consumer products, and despite its known dangers, it is widely used as a building material in the developing world.

ASBESTOS ROCKS

In 2005, some 2.2 million metric tons of asbestos minerals were mined worldwide, primarily in Russia and Canada. The six minerals that are classed as asbestos are chrysotile, amosite, anthophyllite, actinolite, crocidolite and tremolite. These minerals all have a fibrous structure, though the fibres of chrysotile are curly (serpentine) while those of the other five minerals are needle-shaped (amphibole).

The most common asbestos mineral, accounting for 95 per cent of asbestos products, is chrysotile, or 'white' asbestos, which is a mineral obtained from rocks commonly found all over the Earth; amosite, or 'brown' asbestos, is mined in South Africa; and crocidolite, or 'blue' asbestos, is found in both South Africa and Australia. The other three asbestos types are less commonly used in human-made products.

The industrial history of asbestos begins in the nineteenth century with the main phase of the Industrial Revolution. By the middle of the century it was commonly used as an insulation material in both Europe and America; especially chrysotile, the curly fibres of which could be woven into a fabric. By the mid-twentieth century, asbestos was a major material in the building industry, used in fire-retardant coatings; concrete; bricks; pipes and fireplace cement; heat-, fire- and acid-resistant gaskets; pipe insulation; ceiling insulation; fireproof drywall flooring and roofing; lawn furniture and drywall joint compounds. In the 1950s, Kent, the first filter-tipped cigarette, used crocidolite in its 'Micronite' filters, adding the risk of asbestos-related diseases to those of lung cancer, heart disease and bronchitis. And until the 1990s automobile break pads and shoes were made of asbestos, which during use released microscopic asbestos particles into the atmosphere.

DEADLY MINERALS

Unlike other minerals, such as the heavy metals, asbestos is not a poison that attacks the body's tissues and organs directly; it causes disease and death in a completely different way from lead or mercury. Structurally, asbestos consists of molecular lattices, which are extremely fragile. Even the disturbance from day-to-day handling will fracture the fibres into ever-smaller pieces that are so tiny that they are invisible to the naked eye, being much thinner than a human hair. All types of asbestos can cause disease, but the most dangerous is 'blue asbestos', or crocidolite.

A person is at risk if he or she inhales high concentrations of asbestos fibres over a long period, and victims are almost always individuals who have worked directly with the material in mining or manufacture. Family members and others living with asbestos workers also have an increased risk of developing asbestos-related diseases. This is due to exposure to dust brought home on the clothing and hair of asbestos workers. Workers in the building trades who handle and remove asbestos materials during house renovations are also at risk unless they wear protective masks and clothing. However, it is very unlikely that a person would fall ill from a single, high-level dose of asbestos, as would result from disturbing asbestos materials during home improvements.

Asbestos is associated with two serious illnesses: asbestosis and mesothelioma, a form of cancer. The first diagnosis of asbestosis was made in the United Kingdom in 1924. The term 'mesothelioma' was not used in medical literature until 1931, and was not associated with asbestos until the 1940s.

Asbestosis is a chronic inflammatory condition of the lungs caused by scarring by asbestos fibres. It occurs after prolonged exposure to asbestos and is now recognised worldwide as an occupational lung disease. The scarring of lung tissue reduces total lung capacity. Typically, victims suffer from severe shortness of breath, especially during exertion. Coughing is not a typical symptom unless the patient has another respiratory condition. In the most advanced and severe cases the reduction in total lung capacity may induce respiratory or heart failure, leading to death. There is no cure for asbestosis, though breathing pure oxygen alleviates its symptoms.

© Bermau | Dreamstime.com

Mesothelioma is a cancer that is almost always caused by exposure to asbestos. Malignant cells develop in the mesothelium – the protective lining that covers most of the body's internal organs – which includes the pleura (around the lungs), the peritoneum (around the abdominal cavity), the pericardium (heart sac), and the tunica vaginalis (around the testes). Symptoms of mesothelioma may not appear until 20 to 50 years after exposure to asbestos. Shortness of breath, coughing, and pain in the chest due to an accumulation of fluid in the pleural space are often symptoms of pleural mesothelioma. Symptoms of peritoneal mesothelioma include weight loss, abdominal swelling and pain due to a build-up of fluid in the abdominal cavity. Symptoms of pericardial mesothelioma may include blood-clotting abnormalities, anaemia and fever.

REMOVING ASBESTOS
The process of removing asbestos, in this case from an old factory, is painstaking and carries risks for the people performing the task.

In a pattern that we shall find repeated several times in this book, doctors discovered the medical evidence connecting a substance or drug to a debilitating and often fatal disease, but big business and its allies in government succeeded in suppressing it, and opposed legislation for decades, their only motive being greed. The case of asbestos is a classic in this regard, with the first evidence linking asbestos with lung diseases in the developed world coming at the end of the nineteenth century.

The chief inspector of factories in the United Kingdom raised the first doubts about the safety of asbestos in 1898. In 1899, British physician Dr Montague Murray conducted a post mortem on the body of a young asbestos factory worker. He testified that he had found asbestos fibres in the young man's lungs, and that in his opinion they had contributed to or caused his death. Similar cases revealed a link in the United States in the early part of the twentieth century. Further studies in the 1920s and '30s provided overwhelming evidence for the link between asbestos and the disease that became known as asbestosis, and the British government introduced the first safety regulations protecting workers in the industry in the early 1930s.

In the 1940s, the major asbestos manufacturers in the United States realised the dangers to their workforces and sponsored medical research into asbestos-related diseases, but on the condition that the results were kept secret. Internal documentation from the firms reveals that they knew that asbestos causes asbestosis and mesothelioma, and that they kept it secret because of the possible loss of income from lawsuits and environmental regulation. From the 1940s to the 1980s, big business in the United States succeeded in suppressing much of the medical evidence and opposed any legislation controlling asbestos or warning the public of its dangers.

INTERNAL DOCUMENTS SHOWED THAT FIRMS KNEW THAT ASBESTOS CAUSES ASBESTOSIS AND MESOTHELIOMA, BUT KEPT IT SECRET.

However, during the 1980s governments in the developed world finally moved against the importation of asbestos and use of asbestos products, with total bans and removal legislation in the European Union, Japan, Australia and New Zealand. In 1989, the US Environmental Protection Agency finally issued the Asbestos Ban and Phase Out Rule, which was challenged in the courts and overturned in 1991. This means that to this day consumer products in the United States can contain trace amounts of asbestos.

BLACK GOLD: THE USES AND ABUSES OF PETROLEUM

FAILING

Never got off the drawing board

Didn't work in practice

Killed its inventor

A commercial failure

Unforeseen consequences

Was used for evil ends

A success born of failure

Main Culprits: National governments and petroleum majors

Motivation: Greed

Damage Done: Environmental damage on a planetary scale, global warming and oil wars

'In Fuyan County there is shi you (rock oil). I know its soot can be used to make ink stick, which is black and shining, far better than that made of pine wood soot … This thing will surely have wide applications in the world. As the rock oil abounds in the earth, its supply is ample, unlike pine wood, which may someday be exhausted.'

Shen Kuo, 1031–95, *Mengxi Bitan (Dream Pool Essays)*, trans. P. Wang

Walking around our cities, it seems impossible to imagine our civilisation without the use of petroleum and its many products. However, there was nothing inevitable about our current petroleum-based economy, which is a barely more than a century old. Alternatives to petroleum fuels have always existed in the shape of renewables, biofuels and, more recently, hydrogen and compressed air. There are also plenty of biodegradable and recyclable materials that we could use instead of plastics (see pp. 141–5), plastics being the second major product of the petrochemical industry. Had we invested even a fraction of the many trillions of dollars that have been spent on petroleum and its related industries on other less-polluting technologies, we might not now be facing the planetary catastrophe that is global warming, as well as a host of other social, economic and environmental problems. However, a combination of greed and short-termism on the part of big business and corruption and a lack of will on the part of governments allowed a limited, non-renewable resource to become the main energy source for the planet and the raw material for many of our consumer goods.

ROCK OIL | The ancients used petroleum, known to the Chinese as *shi you*, 'rock oil', in a variety of ways. In ancient Babylon, a major city-state of Mesopotamia, 'the land between the two rivers', (now Iraq), builders used asphalt to waterproof their massive brick ziggurats, palaces and city walls, which were at constant risk from flooding. The ancient Persians used rock oil both as a medicine and a lighting fuel. The Romans obtained rock oil from the province of Dacia (modern-day Romania) for their oil lamps.

The first oil wells were drilled in China in the fourth century BCE, using bamboo pipes and drill bits to extract and transport the oil. In both China and Japan, rock oil was used for lighting and heating, as well as evaporating seawater to produce salt. In Byzantium, petroleum had a military application in the shape of 'Greek fire', an inflammable substance fired under pressure from ships that saved the city from invasion many times between the seventh and the thirteenth centuries CE. In the Middle East, tar from petroleum deposits was used to surface the streets of Baghdad in the eighth century. One hundred years later, a Persian chemist distilled a kerosene-like substance from petroleum, which was used in oil lamps.

The history of the modern petroleum industry begins in 1852, when the Polish chemist Ignacy Lukasiewicz (1822–82) first refined the fuel oil that would later be known as kerosene from natural petroleum. The first petroleum 'mine' opened in southeast Poland in 1853. A year later, a Yale chemistry professor, Benjamin Silliman (1779–1864), succeeded in fractionating petroleum by distillation, and the petrochemical industry was on its way.

FROM KEROSENE TO GASOLINE

In 1859, the first American oil well was sunk at Oil Creek, Pennsylvania. It initially yielded a paltry 25 barrels a day. Until the 1880s, petroleum was principally in demand for kerosene for lighting, and it is just conceivable that, after the introduction of electricity for lighting, petroleum-based products would have disappeared like so many other outdated technologies. What saved the nascent oil industry was the development of another invention, the internal combustion engine (see pp. 108–12), which created an unprecedented demand for petroleum and its derivatives – diesel, gasoline and engine oil – which is still with us today. From the turn of the century, the voracious requirements of the automobile industry ensured that petroleum, which was then plentiful and cheap to extract, would be preferred over less polluting but more expensive early biofuels.

Production of petroleum in the United States increased from around 2,000 barrels in 1859 to more than 126 million barrels in 1906. By 1910, significant petroleum finds had been made in North America, the Far and Middle East, and Central and South America. The current top three oil-producing nations are the kingdom of Saudi Arabia, Russia and the United States. About 80 per cent of the world's accessible reserves are located in the Middle East, with 62.5 per cent in the Arabian Peninsula alone. Petroleum accounts for a large percentage of the world's energy consumption, ranging from 32 per cent in Europe and Asia to 53 per cent in the Middle East. Rates of petroleum dependence in other regions are 44 per cent in South and Central America, 41 per cent in Africa, and 40 per cent in North America. Despite the development of alternative fuels, such as hydrogen and biofuels, about 90 per cent of vehicle needs are met by petroleum.

The world consumes 30 billion barrels of crude oil per year; the top consumers are the developed nations, with almost a quarter of the oil consumed going to the United States, followed by Russia, China and India. In net dollar worth, the production, distribution, refining and retailing of petroleum products represent the single largest industry on the planet.

Although the Western world is completely dependent on petroleum, it cannot even begin to produce enough to meet its own needs. Moreover, the remaining reserves are largely located in regions that are either considered unfriendly to the West or politically unstable, thereby threatening the continuity of supply. This was made clear in the 1973 'oil shock', the 1979 'energy crisis', as well as the 2008 hike in the price of crude that saw it reach close to US $150 a barrel.

OIL INSECURITY

Oil insecurity has been indicted as one of the main causes for the recent wars in the Middle East: the Iran–Iraq war of the 1980s, the Gulf War in the early '90s, and the Iraq War a decade later. By extension, disputes over oil can be seen as one of the major causes of the radicalisation of Islam, in particular in Iran, where British and American intervention over the country's huge oil reserves in the 1950s, which installed the autocratic regime of the Shah, led to the Islamic revolution of 1978. The invasion of 2003 Iraq, also criticised by many as an 'oil war', is cited by Al-Qaeda as one of the main reasons for its campaign of terror against the West.

In addition to the twin problems of supply insecurity and oil wars, there are grave doubts about the remaining reserves of petroleum. 'Peak oil' theory states that oil discoveries worldwide peaked in 1965. Supporters of the theory argue that the fact that production has exceeded discoveries in every year since 1980 proves that they are right. It is now generally accepted that US petroleum production peaked in 1971, although the debate continues as to whether world production peaked in 1965, 1989, 1995 or 2000. What analysts do agree on, however, is that production has already peaked, and that we are on the slow but inexorable road to running out of our favourite energy fix.

On the global scale, however, war, terrorism and peak oil can be considered the relatively minor consequences of our dependence on petroleum as the main energy source for our civilisation. The far greater threat to the human race is the damage that petroleum consumption has done to the environment. This comes both in the form of one-off events that degrade the environment, such as the localised damage caused by extraction and marine spills; and on the planetary scale, with the growing impact of global warming and climate change caused by the burning of the fossil fuels: oil, coal and gas.

As the large reservoirs of crude oil are exhausted, the search is on for alternative and non-conventional petroleum sources. The greatest of these are tar sands (bituminous sands) in which petroleum is mixed with clay and sand. The largest deposits are in Canada and Venezuela, and reserves held in tar sands are thought to equal estimated conventional crude oil reserves. The problem, however, is that their extraction is extremely damaging to the environment, as the sands have to be strip-mined. The extraction of petroleum from the sands is also costly in terms of water and energy; therefore, to make the exploitation of tar sands worthwhile would require a barrel price of US $80–100.

THE LARGEST TAR SAND DEPOSITS ARE IN CANADA AND VENEZUELA, AND RESERVES ARE THOUGHT TO EQUAL ESTIMATED CONVENTIONAL CRUDE OIL RESERVES.

Although the exploitation of tar sands may put off the inevitable exhaustion of petroleum, it may cripple the world economically. Moreover, another problem with petroleum from tar sands is that it emits more greenhouse gases, which is the most dramatic effect of our petroleum addiction.

Although a few scientists and politicians continue to dispute the evidence for human-caused climate change, the case is now compelling. The concentration of carbon dioxide (CO_2) in the atmosphere has increased by a third since the beginning of the Industrial Revolution in the mid-1700s. These levels are considerably higher than at any time during the past 650,000 years, according to climate data extracted from ice cores.

The burning of fossil fuels has produced approximately three-quarters of the increase in CO_2 due to human activity over the past two decades. Oil produces 15 per cent less CO_2 than coal, but 30 per cent more than natural gas. However, the unique role of petroleum as the main source of transportation fuel makes reducing its CO_2 emissions a real challenge. Although large power plants can eliminate their CO_2 emissions by using techniques such as 'carbon capture', these techniques will not work for individual automobiles.

The current atmospheric concentration of CO_2 is about 385 parts per million (ppm). Future CO_2 levels are expected to rise due to the continued burning of fossil fuels. The rate of increase will depend on uncertain economic, social, technological and natural developments, but may be ultimately limited by the availability of petroleum and other fossil fuels.

The Intergovernmental Panel on Climate Change has outlined a range of possible scenarios for the increase in atmospheric CO_2, ranging from 541 to 970 ppm by the year 2100. Unfortunately for the planet, the reserves of fossil fuels are quite sufficient to reach this level with the disastrous consequences that it entails for the climate and sea levels.

THE WORLD'S HUNGER FOR OIL

Based on data from the 2008 *CIA World Factbook*, this map shows the consumption of oil by the nations of the world, in billions of barrels per day.

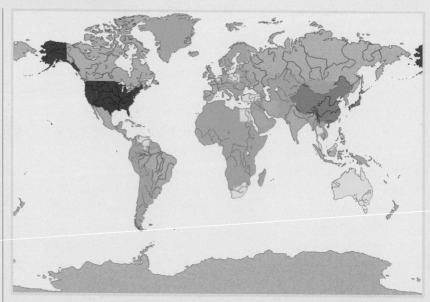

- > 10 billion barrels/day
- 5 > 10 billion barrels/day
- 1 > 5 billion barrels/day
- 0.5 > 1 billion barrels/day
- < 0.5 billion barrels/day
- Data not available

DEADLY MATTERS: THE ORIGINS AND DEVELOPMENT OF CAPITAL PUNISHMENT

FAILING

Never got off the drawing board

Didn't work in practice

Killed its inventor

A commercial failure

Unforeseen consequences

Was used for evil ends

A success born of failure

Main Culprits: Rulers throughout the ages

Motivation: Punishment and retribution

Damage Done: Inhumane punishment and the deaths of thousands of innocents

'And Harbonah, one of the chamberlains, said before the king, "Behold also, the gallows fifty cubits high, which Haman had made for Mordecai, who had spoken good for the king, standeth in the house of Haman." Then the king said, "Hang him thereon." So they hanged Haman on the gallows that he had prepared for Mordecai. Then was the king's wrath pacified.'

Book of Esther, 7:9–10, King James Bible

Humans have always shown considerable ingenuity in devising ways of dispatching those they find guilty of crimes so heinous that they deserve to die – be it for the taking of another life, the worship of the wrong god, or the theft of a loaf of bread. Methods of judicial execution have included: hanging; crucifixion; burning at the stake; impalement; drowning; breaking on the wheel; cudgelling; being torn apart, mauled or trampled by animals; disembowelling; decapitation with sword, axe or guillotine; garroting; electrocution; firing squad; gassing and lethal injection.

Objections to the death penalty do not only rest on the argument that in most of its forms it constitutes a 'cruel and unusual' punishment, but also on the fact that there have been many wrongful convictions that have been subsequently overturned. However, once the sentence has been carried out, it is rather too late to say sorry.

PLAYING HANGMAN

According to the Book of Esther, Haman, vizier to King Ahasuerus of Persia (possibly Artaxerxes I, 465–424 BCE) invented the first gallows. Angry because the Israelite Mordecai had refused to pay homage to him, Haman plotted to kill all the Jews in Persia. To this end he built a gallows upon which to hang Mordecai, and then denounced him to the king. Queen Esther, Ahasuerus's Jewish wife, foiled the plot, however, and it was Haman and his ten sons who were executed on the gallows.

Despite these rather unfortunate origins for its inventor, hanging has remained a popular mode of execution from ancient times to the present day. For alfresco executions in the country, a convenient tree would suffice; in towns, a gallows had to be set up temporarily or permanently for the edification of spectators. The oldest and simplest form of gallows was a gibbet with a vertical post and a horizontal beam to which the rope was attached. During public hangings, the victim would die slowly by strangulation, and the body would usually be left as a warning to would-be wrongdoers.

With the increase in population and in the number of offenses liable to the death penalty – at one point 222 crimes merited the death penalty in England – the authorities were obliged to create multi-user gallows. The triangular gallows at London's public execution ground of Tyburn Field, for example, could accommodate 24 victims at one time. In more humane times, when executions took place more privately inside prisons, the gallows were erected on a scaffold equipped with a trapdoor. The victim dropped through the trapdoor and died quickly from a broken neck, rather than slowly from strangulation.

Apparently, however, humanity remained dissatisfied and devised and practised ever more barbaric methods of execution with relish during the Middle Ages and Renaissance. One infamous example, brought to the screen by Mel Gibson in the movie *Braveheart*, was the execution of the rebel (if you are English) or freedom fighter (if you are Scots) William Wallace (c. 1272–1305) in London. After his trial, Wallace was dragged through the streets naked and taken to the execution ground at Smithfield. He was 'hanged, drawn and quartered' – that is, hanged on a gallows but not to death; castrated and disembowelled, with his genitals and innards burned before him; and finally beheaded. As a finale, the corpse was cut into four pieces, and the severed head exhibited on a spike on London Bridge.

With the dawning of the Age of Enlightenment in the eighteenth century, reformers attempted to make executions more humane and more 'democratic'. In France, before the revolution of 1789, executions were particularly barbaric. Criminals were burned at the stake or 'broken' on a wheel with a hammer or an iron bar. If they were not killed by their injuries, they were left on the wheel to die of thirst. Executions also varied with the class of the victim. The rich could afford to bribe the executioner to give them a quick death by decapitation, while the poor were often subject to a slow death by hanging.

These cruel and unusual forms of punishment appalled Joseph-Ignace Guillotin (1738–1814), a doctor and member of the first French National Assembly. In a debate on the death penalty, he proposed that henceforth criminals should be decapitated by means of a 'simple mechanism' that would behead them quickly and painlessly.

Guillotin was in reality an abolitionist, and he hoped that the introduction of a more humane means of execution would be a first step to ending capital punishment altogether. Although he played little part in the development of the guillotine itself, his advocacy of it was sufficient to associate his name with it forever. 'Madame Guillotine' was

THE RICH COULD AFFORD TO BRIBE THE EXECUTIONER TO GIVE THEM A QUICK DEATH BY DECAPITATION, WHILE THE POOR WERE OFTEN SUBJECT TO A SLOW DEATH BY HANGING.

not a completely novel invention, but was based on earlier Italian and Scottish machines that crushed the necks of their victims or beheaded them by brute force. The guillotine, in contrast, used a hinged yoke to hold the victim in place and a slanted blade to sever the head cleanly from the body. Guillotin very nearly became a victim of the mechanism he had advocated when he fell from favour during the revolutionary Reign of Terror of 1793–4. Fortunately, before he could be executed, the dictator Robespierre was deposed and himself guillotined, and the good doctor was spared.

Another intended 'humane' method of execution was the electric chair. In 1887, after a particularly gruesome hanging, the State of New York decided to find a less barbaric humane form of execution. One suggestion, from the dentist Alfred Southwick, was to electrocute the victim while he or she was secured to a chair.

At that time electricity was then rapidly replacing gas and kerosene for lighting, and the electric chair played its part in the 'war of currents' – the bitter rivalry between Thomas Edison, promoter of direct current (DC) on one side, and George Westinghouse and Nikola Tesla (see pp. 156–9), who had invented the alternating current (AC) on the other. Neither side wanted his invention to be associated with capital punishment, but it was Edison who mounted the most successful and underhand campaign against his rivals. He claimed, falsely, that AC was more lethal than DC, and arranged the electrocution of animals for the press. When Westinghouse refused to supply an AC generator for the first electric chair, Edison bought one anonymously, claiming that it was destined for a university.

© Public Domain

JOSEPH-IGNACE GUILLOTIN
The man credited by history with inventing the guillotine during the French Revolution almost became one of its victims.

The first person to go to the electric chair was William Kemmler, who was executed on August 6, 1890. For all its vaunted modernity and humaneness, its first use was as gruesome as it was farcical. The initial 17-second shock caused the condemned man to black out, but did not kill him. The doctors present suggested an immediate second shock, but the generator needed several minutes to build up a sufficient charge. At the second attempt Kemmler received a 2,000-volt shock. He died from heart failure, but such was the intensity of the shock that his blood vessels burst, causing massive internal and external bleeding, and his body caught fire. Westinghouse reportedly commented, 'They would have done better using an axe.'

The chair has been used since in 25 states of the Union, and by one foreign country, the Philippines, between 1924 and 1976. Despite its many drawbacks, the chair remained the preferred method of execution in many states until the mid-1980s, when it was replaced by lethal injection. It remains an elective method in certain states. The last elective use of the electric chair at the time of writing was the execution of James Earl Reed on June 20, 2008, in South Carolina.

In addition to being considered by many to be a cruel and unusual punishment, the death penalty is, of course, irreversible. There have been 39 known cases in the United States in which executions have been carried out despite serious doubt over the guilt of the victim. In the United Kingdom, four people who were executed between 1950 and 1953 were posthumously exonerated or pardoned. Although many countries in the world have abolished or suspended the death penalty, it is still practised in certain countries, notably the United States, China and Iran. In the latter two countries, the death penalty is sometimes imposed for acts that would not be recognised as crimes in other parts of the world, such as the execution of men in Iran for being homosexual, and the execution of political dissidents in China. In these circumstances, the death penalty is not an instrument of justice but of unjust state oppression.

TOO LATE TO SAY SORRY

FAILING

Never got off the drawing board

Didn't work in practice

Killed its inventor

A commercial failure

Unforeseen consequences

Was used for evil ends

A success born of failure

GOING UP IN SMOKE: NATIVE AMERICANS INVENT SMOKING

Main Culprits: Tobacco growers and manufacturers as well as national governments

Motivation: Greed

Damage Done: Untold millions of deaths from lung cancer and heart disease

'It makes a man sober that was drunke. It refreshes a weary man, and yet makes a man hungry. Being taken when they goe to bed, it makes one sleepe soundly, and yet being taken when a man is sleepie and drowsie, it will, as they say, awake his braine, and quicken his understanding.'

King James I of England, from *A Counterblaste to Tobacco*, 1604

The native peoples of North and South America were the first to smoke the dried leaves of plants of the genus *Nicotiana*. In extremely high doses, tobacco is a hallucinogen, and it is thought that Native American priests and shamans first used tobacco to induce trances during which they communicated with their gods, spirits and ancestors. Recreational smoking spread to the elite of the Maya in their classical era (c. 900 CE), and they are shown in temple and palace bas-reliefs using 'smoking tubes'. The Aztecs who dominated Central America until the Spanish Conquest in the sixteenth century also used tobacco in their religious ceremonies and during social occasions. The goddess Cihuacoatl was the divine embodiment of tobacco, and her priests wore tobacco gourds during her ceremonies, which included the most gruesome forms of human sacrifice. Tobacco, it seems, claimed its first human victims long before the advent of cigarettes.

THE GODDESS CIHUACOATL WAS THE DIVINE EMBODIMENT OF TOBACCO, AND HER PRIESTS WORE TOBACCO GOURDS DURING HER CEREMONIES.

By the time the Spanish Conquistadors arrived in the Americas in the early sixteenth century, the recreational smoking of tobacco was widespread throughout the region. At Aztec banquets, the meal would start with the distribution of flowers and smoking tubes to the guests. At the end of the meal, the remaining smoking tubes would be distributed as alms to the servants, the old and the poor. In the Caribbean, Mexico, and Central and South America, early forms of cigarettes, such as smoking reeds, tubes or cigars, were the most common means of smoking. In North America, however, the most common smoking implement was the pipe, which has been immortalised in thousands of Westerns as the 'peace pipe' that was offered to European settlers as a gesture of goodwill.

Within a century of the colonisation of the Americas by Europeans, the smoking, cultivation and trading of tobacco had spread to every continent. Tobacco, both in the form of dried leaves and the plants themselves, followed the trade routes to ports, markets and major towns, and from there into the countryside. By the mid-seventeenth century every major country had imported smoking, and in many cases had assimilated it into its own culture, in spite of the attempts of rulers to stamp out the practice with harsh penalties and fines. When smoking was introduced to Europe, the concept of ingesting substances in the form of smoke was largely unknown. The first report of smoking in England is of a sailor in Bristol in 1556 seen 'emitting smoke from his nostrils'. The English term 'smoking' was coined in the late seventeenth century, and until then it was referred to as 'drinking smoke'. The Frenchman Jean Nicot (to whom we owe the word 'nicotine') introduced tobacco to France in 1560. Like tea, coffee and opium, tobacco was one of the many substances that were originally thought to have medicinal uses. Early European medicine was based on the ancient Greek theory of the humours – the idea that everything had a specific nature that could be hot or cold, and dry or moist. Tobacco was seen as a substance that was both heating and drying, and accordingly doctors assigned it an endless list of beneficial effects.

'A HEINOUS CRIME' The spread of tobacco through the world, although staggeringly fast, did not go unopposed. Civil and religious leaders in both the East and West fulminated against the practice and tried to ban it. Sultan Murad IV (1612–40), ruler of the Ottoman Empire, was among the first to attempt a smoking ban by claiming it was a threat to public morals and health. The Chinese emperor Chongzhen (1611–44) issued an edict banning smoking two years before the overthrow of the Ming dynasty. The rulers of the Qing dynasty, who were of hardy nomadic stock, proclaimed smoking to be 'a more heinous crime than neglecting archery'. In seventeenth-century Japan, the Shogun condemned the growing of tobacco as a threat to the national economy by using up valuable farmland that could be used for growing rice. James I of England (1566–1625), a staunch anti-smoker and the author of *A Counterblaste to Tobacco*, tried to curb the new fashion by imposing a staggering 4,000 per cent excise duty on tobacco sales in 1604, but the move proved a complete failure. At the time London already had 7,000 tobacco sellers. In 1634, the Patriarch of Moscow forbade the sale of tobacco and sentenced men and women who broke the ban to have their nostrils slit and to be whipped until the skin was flayed off their backs. Pope Urban VIII also condemned the practice of smoking in a papal bull of 1642.

Although at first tobacco was seen as an expensive luxury or a medicine, it quickly became an accepted part of daily life among men and, later, women and adolescents of both sexes. In the seventeenth century, Western navies incorporated a tobacco allowance in their standard naval rations. During World War I (1914–18), cigarette manufacturers and governments collaborated in securing tobacco and cigarette supplies for the soldiers fighting in the trenches.

Until the mid-twentieth century, the majority of the adult population in many Western countries smoked, and the warnings of its ill effects were ridiculed and ignored. Today, the anti-smoking movement has considerably more weight and evidence of its claims, but a considerable proportion of the population remains steadfast smokers. It is estimated that around 1.1 billion people still smoke worldwide.

The image of the smoker varies considerably, but smoking is often associated, especially in novels and movies, with individuality and aloofness. However, smoking also reinforces existing social practices and is part of the rituals of many social groups. Smokers often start smoking in social settings, and the offering and sharing of a cigarette is often an important rite of initiation or simply a good excuse to start a conversation with a stranger in a public space. Lighting a cigarette is also often seen as an effective way of looking busy. For adolescents, it can be the first step out of childhood or an act of rebellion against the adult world. The rise of the modern anti-smoking movement has done more than create an awareness of the dangers of smoking; it has also provoked a reaction on the part of smokers who perceive the bans as an assault on their personal freedom. To a degree this has reinforced a social identity among smokers as rebels and outcasts, who are a persecuted minority set apart from non-smokers.

© Public Domain

NICOTIANA TABACUM
Illustration by Franz Eugen Köhler, from *Köhler's Medicinal Plants* (1887).

During smoking, the main pathway for toxins entering the body is inhalation into the lungs. The incomplete combustion produced by burning tobacco produces carbon monoxide, which impairs the ability of blood to carry oxygen. There are several other toxic compounds in tobacco that constitute serious health risks to long-term smokers, causing conditions that include lung cancer, coronary heart disease, stroke, impotence and low birth weight of infants born of mothers who smoke. Tobacco-related diseases are some of the biggest killers in the world and one of the biggest causes of premature death in the developed world. In the United States approximately half a million deaths per year are due to smoking, while a recent study estimated that one-third of Chinese men will have shortened lifespans because of smoking. Although Dr Benjamin Rush claimed that tobacco was bad for human health as early as 1798, it was only in the twentieth century that the first serious studies into the effects of smoking were conducted. One of the landmark studies came in 1948, when British physician Sir Richard Doll published research that proved that smoking could have adverse effects on human health. However, for decades governments and tobacco companies colluded in suppressing the information.

Inhaling the vaporised gas form of tobacco into the lungs is a quick and effective way of delivering the chemical substances contained in smoke into the smoker's bloodstream and to his or her brain. Nicotine, which is present in tobacco, is known to be highly addictive. Hence, regular smokers quickly become habituated to the presence of nicotine in their bloodstream, and will crave a cigarette if the nicotine level drops. Therefore modern anti-smoking therapies use nicotine replacement gums and patches to gradually reduce the body's dependence on nicotine, and hence manage the craving for cigarettes.

THE INCIDENCE OF TOBACCO-RELATED ILLNESSES

A map showing the prevalence of current tobacco use in people aged over fifteen. Based on 2005 data from the World Health Organization Statistical Information System (www.who.int/whosis).

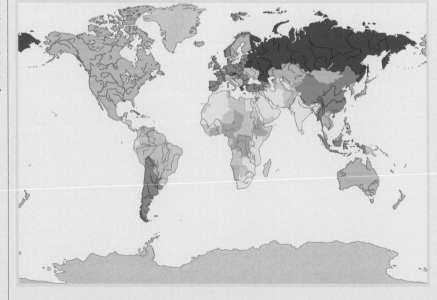

- > 40%
- 30–40%
- 20–30%
- 10–20%
- < 10%
- Data not available

In recent decades governments in the developed world have been trying to deter people with campaigns in the mass media that stress the harmful long-term effects of smoking. Passive, or second-hand, smoking is one of the major reasons given for the enforcement of bans in public places. The idea behind this is to discourage smoking by making it more inconvenient, and to stop harmful smoke being released within enclosed public spaces where non-smokers, and especially the people who work there, are exposed to it.

Although nicotine is a highly addictive drug, its effects on reasoning are not as strong, noticeable or debilitating as those of other drugs such as cannabis or alcohol. Hence governments do not feel there is a justification for a total ban on smoking. However, social pressure against the habit is so strong in some parts of the developed world that, for all intents and purposes, they have become smoke-free zones.

Tobacco legislation remains a hot topic, however, and in 2003, after four years of research and negotiation, the World Health Organization created the Framework Convention on Tobacco Control 'in response to the globalisation of the tobacco epidemic'. The convention entered into force in 2005, and has since become one of the most widely embraced treaties in United Nations history, with 168 signatories. It focuses on reducing both the supply of and the demand for tobacco products, and ensures that tobacco use will remain a controversial subject for years to come.

SMOKING BANS

FAILING

Never got off the drawing board

Didn't work in practice

Killed its inventor

A commercial failure

Unforeseen consequences

Was used for evil ends

A success born of failure

BANG! BANG! YOU'RE DEAD: THE INVENTION OF GUNPOWDER AND GUNS

Main Culprits: Daoist alchemists

Motivation: The desire for immortality

Damage Done: Slaughter on an unprecedented scale

'The barrel is made of iron, three feet (1 m) long, with a handle two feet (60 cm) long and the weapon is used by foot soldiers. It has a range of 300 paces. The enemy can be shot with pellets or struck with the gun itself.'

Description of a Chinese 'fire-lance' quoted in *Gunpowder*, C. Ponting, 2005

Undoubtedly one of the great ironies of history is that 'black powder', known to us as gunpowder, was invented by Chinese alchemists who were searching for the secret of immortality – the elixir of eternal life. The ancient Chinese Daoists had a quite different conception of immortality from Christians, for whom the body perishes but the soul lives on. In Daoism, there is no division between body and soul, matter and spirit; therefore, for someone to attain immortality, both body and spirit have to be preserved. To achieve this magical state of preservation, alchemists searched for the elusive and mysterious concoctions of elements that would make the human body incorruptible. In the process, they created the science of experimental chemistry, and discovered many of the basic chemical elements and compounds, including the three constituents of gunpowder: carbon (C), sulphur (S) and saltpetre (potassium nitrate, KNO_3).

The Chinese perfected the recipe for a proto-gunpowder by the ninth century CE. At this time, the mixture, which contained a low proportion of saltpetre, did not explode but burned fiercely. Therefore, its first use as a weapon of war was as an incendiary added to resins, oils and dried plant matter, and launched by catapult in firebombs against enemy armies, ships and cities. In the next two centuries, the Chinese developed a range of weapons that exploited the inflammability of black powder. One of the first of these inventions was the incendiary arrow tipped with gunpowder, which was launched singly and then in volleys from the first multiple-launcher weapon in history. Around 950 CE, the Chinese invented the 'fire-lance' – at first, a paper container strapped to a lance that once ignited became a flamethrower with a range of 12 ft (3.6 m). The fire-lance is the ancestor of all firearms, as later models of the weapon were made with a metal barrel (see quote). It could be packed with projectiles, such as fragments of pottery, shards of metal, sand or arsenic balls, which would be ejected with the flame when the lance was ignited. These weapons, however, were not yet true guns but more correctly termed 'eruptors'.

The first weapon to make use of the explosive nature of gunpowder was the 'thunderclap bomb', which the Chinese began to manufacture around 1000 CE. The gunpowder was packed inside a bamboo and paper shell; the bomb did little actual damage but made a loud noise

FIRE AND BRIMSTONE

and created smoke that disoriented enemy soldiers – especially the nomadic tribesmen that periodically attacked China, who would not have been used to such explosions. Two hundred years later, however, the Chinese had improved on its design and killing power by packing gunpowder and metal shards inside a metal casing. However, like all early gunpowder weapons, it was unreliable as it had to be ignited before being launched by catapult and could easily misfire or explode too early. Nevertheless, the thunderclap bomb was an important step in developing more explosive forms of gunpowder that would be needed to launch projectiles at speed over much greater distances.

With more powerful high-saltpetre gunpowder and improvements in metal casting, the stage was set for the development of the next generation of gunpowder weapons, which included the first true guns (the definition of a gun being a weapon that fires a single projectile that is almost the same size as the barrel). The first cannons were manufactured in China in the twelfth century, and were made first of bronze and later of stronger cast-iron. The earliest designs were vase-shaped and had extremely thick walls so as not to be blown apart by the charge needed to fire the projectile. The gunpowder and projectile were loaded from the front, and there was a touchhole to light the gunpowder at the rear.

THE EARLIEST HANDHELD GUN DISCOVERED TO DATE, AND THUS THE ANCESTOR OF EVERYTHING FROM THE FLINTLOCK PISTOL TO THE .44 MAGNUM, WAS MADE AROUND 1285 CE.

The earliest handheld gun discovered to date, and thus the ancestor of everything from the flintlock pistol to the .44 Magnum, was made around 1285 CE. The weapon was made of bronze, was 1 ft (30 cm) long, and weighed 8 lb (3.6 kg). Like the early cannons, it was muzzle-loaded with a touchhole at the rear. As a weapon it was probably extremely inefficient and unreliable. The gunner had the difficult task of aiming and lighting the charge with a taper through the touchhole at the same time. If the gun did not explode in the gunner's hands, it often misfired or failed to fire, because a sudden shower put out the taper or made the gunpowder too damp to ignite. Even if it did go off, the projectile's range and flight path were erratic. The only sure way to hit your enemy was if he was standing still right in front of you. At that range, however, the heavy metal gun would have been just as effective as a club. Finally, the weapon took many minutes to reload, during which time the gunner was vulnerable to attack.

However, these early weapons proved effective in two ways. First, if used in numbers, they could inflict casualties on the massed ranks of a charging army, as the projectiles could pierce any armour then in use. Second, when used against troops who were unfamiliar with gunpowder weapons, the noise and smoke they emitted caused both men and horses to panic.

GO WEST

The Chinese attempted to keep the secret of gunpowder and gunpowder weapons to themselves, but even their arsenals of fire arrows, bombs, landmines, rockets, fire-lances, cannons and guns were not sufficient to protect them from the waves of nomadic invaders that regularly attacked the country. The first of these to obtain gunpowder weapons were the nomadic Jurchen, who invaded in the twelfth century. But the true transmitters of gunpowder technology to the world were the Mongols, who defeated the Chinese Sung in 1276. Within 50 years, the Mongols had established an empire that stretched from China to Europe. The Mongols took primitive gunpowder weapons with them in their relentless westward expansion through the Islamic empires of Iran and the Middle East. By the early 1300s, Islamic armies had fire-lances; by the 1320s, they had developed primitive artillery. It was only a matter of time before the technology reached Western Europe.

Although gunpowder came to Europe late, European armies wasted no time in adopting the new weapons. By the late fourteenth century, cannon had replaced siege engines in siege warfare, and by the early fifteenth, the first handgun, the German-made *hackenbuchse* (hook gun), was in service. The *hackenbuchse* was similar to the earlier Chinese guns, being muzzle-loaded and lit from a touchhole. It also suffered from the same drawbacks: it was unreliable, with poor aim and range, a tendency to explode, and was very slow to reload. However, by the mid-sixteenth century, the gun had made most other battlefield weapons redundant, with the exception of the pike, which was used to protect gunners from infantry and cavalry charges while they reloaded.

European craftsmen quickly improved on earlier Chinese and Islamic models. The muzzle-loaded matchlock musket was invented in Germany in 1475. It solved the problem of aiming and priming at the same time, but it was cumbersome and heavy, and firing relied upon a match that had to be kept alight, even in wet conditions. One expression we have inherited from the matchlock gunner is a 'flash in the pan', which occurred when the powder in the pan caught fire but failed to trigger the main charge that fired the bullet.

The matchlock was replaced in the early 1500s by the slightly better wheel-lock musket, which in turn gave way to the flintlock 50 years later. The flintlock was much more dependable; the spark used to ignite the gunpowder was supplied by a sharpened piece of flint clamped in the jaws of a 'cock', which, when released by the trigger, struck a piece of steel. Although the flintlock was lighter and more reliable, it still had a firing rate of just one shot per minute. Nevertheless, the flintlock remained the weapon of choice for the military until the mid-nineteenth century.

GREAT BALLS OF FIRE

At the same time as handheld firearms began to dominate the medieval battlefield, gunpowder weapons replaced the catapults that had been in use since Roman times to attack enemy emplacements and fortifications. The earliest European cannons, known as bombards, were developed in the thirteenth century but came into their own in the fifteenth. Bombards grew into massive siege weapons, used to batter the walls of castles and cities. Like other early firearms, muzzle-loaded bombards were prone to misfiring and often exploded, killing the artillerymen who operated them. Moreover, early pieces of ordnance were so heavy and unwieldy that once they had been set in place they were effectively immobile, making them and the soldiers who manned them vulnerable to attack from the defenders.

One of the largest and most famous cannons of the Middle Ages was Sultan Mehmed II's giant bombard, with which he demolished the walls of Constantinople in 1453. This monster had to be cast in the field, weighed 19 tons, and required 200 men and a team of 60 oxen to move into position in front of the formidable double wall of the city that had never been breached in a thousand years. The seventeenth century was the heyday of stationary muzzle-loaded cannons, which were mounted on land fortifications and the decks of warships. In the late eighteenth century, the French perfected the first truly mobile field guns, fired by a weatherproof flintlock mechanism and drawn by a gun carriage, during the revolutionary period and the Napoleonic Wars (1792–1815).

The era of gunpowder weapons lasted for almost one thousand years, from the invention of black powder in China and its first applications in warfare, to the late nineteenth century when it was replaced in firearms and artillery by high explosives (see pp. 118–22). In that time, the Europeans, who were the last to obtain the technology, had made it their own. Until the sixteenth century, China, India and the Islamic world had led the world in gunpowder technology. After the seventeenth they were left behind, and by the nineteenth the British had turned it against its inventors in the two devastating military defeats of the First and Second Opium Wars (1839–43 and 1856–60).

But what allowed Britain, at that point a technological backwater that was poor in natural resources and technological expertise, to overtake and defeat the greatest empires of the Middle Ages in a matter of centuries? The explanation may lie in the very success and stability of Mughal India and Imperial China. While India and China enjoyed centuries of relative unity and peace under centralised governments, Europe was divided into hundreds of petty rival kingdoms, principalities, duchies, republics and city-states that were constantly at war with one another. In the year 1500 there were some 500 states between the Atlantic and the Caucasus, but by the end of the nineteenth century these had been whittled down to a few dozen. But even in the era of the nation states, the wars did not stop.

SURRENDERING ARMS
German soldiers surrender at the Danish–German border at the conclusion of World War II; they discard their firearms in a pile as they file by.

The European powers continued to fight among themselves, and when they had fought themselves to a standstill after the Napoleonic Wars, they turned their attention to the rest of the world. Constant warfare, it is sad to say, drove the economic and technological engines of Europe from the Renaissance to the end of World War II (1939–45). The production of gunpowder-based weaponry stimulated the development of chemistry and metallurgy, two of the industries that would give the West supremacy during the Industrial Revolution.

INHERITANCE POWDER: ARSENIC, THE POISONER'S FAVOURITE

Main Culprits: Alchemists

Motivation: Greed

Damage Done: The murder of thousands

'Again it is not possible to know exactly how many people Mary Ann Cotton murdered, but they almost certainly included her mother, three husbands, a lover, eight of her own children, and seven stepchildren, making a grand total of 20.'

John Emsley, *Elements of Murder*, 2005, on one of the nineteenth century's most notorious arsenic poisoners, Mary Ann Cotton (1832–73)

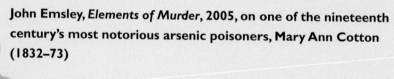

Arsenic is a naturally occurring chemical element (number 33 in the periodic table, chemical symbol: As) found in both organic and non-organic forms. Although in the popular imagination it is best known as a poison, arsenic is widely found in the environment: in soil; water; plants and animals, especially fish and shellfish; and in minute quantities in the human body, where it plays a vital role in metabolism. Arsenic is also used in many industrial processes, and was used as a common colouring agent until the nineteenth century. As with many poisons, in minute amounts arsenic acts as a tonic, hence its use since antiquity as a medicine and tonic and more recently as a doping agent for racehorses. However, if ingested by accident or administered by design, it can lead to chronic poisoning, and in sufficient quantities, it will lead to acute illness and death.

The ancient Chinese, Indians and Romans knew of various forms of naturally occurring arsenic and used it in their medical remedies; it is possible that they also used it as a poison. The form of arsenic that later became known as 'inheritance powder', because of its use by murderers to dispose of inconvenient relatives to inherit their property, is arsenic trioxide (As_2O_3). The Arab alchemist Abu Musa Jabir ibn Hayyan (c. 721–815), known under the Latinised name of Geber in the West, was the first to prepare this white, tasteless and odourless form of arsenic, while the European alchemist Albertus Magnus (1193–1280) succeeded in isolating the element in around 1250. Arsenic trioxide quickly became the murderer's favourite poison because it was impossible for the victim to taste it, and because the symptoms of arsenic poisoning were similar to those of food poisoning and cholera, so doctors were unlikely to recognise it. It was only in the early nineteenth century that a reliable test for arsenical poisoning was discovered, but until the end of the century, because mortality was high from cholera and other intestinal diseases with the same symptoms, it is probable that many instances of poisoning were never correctly diagnosed.

INHERITANCE POWDER

Arsenic can enter the body through the skin, lungs or stomach lining, though the latter is the only realistic way of poisoning someone intentionally. The fatal dose varies between individuals, depending on their size and state of health. As little as 125 mg has been known to kill an adult, but the usual dose is 250 mg. Unlike other toxic substances, such as mercury, arsenic does not remain in the body, so a victim in good health can expel it and recover completely from a non-fatal dose. A dose insufficient to kill will produce symptoms similar to those of food poisoning; hence, amateur poisoners were able to experiment with dosages without fear of detection until their victim finally succumbed.

The symptoms of a fatal arsenic poisoning vary, but often start with a mild headache. The body's attempt to expel the poison from the body triggers vomiting and diarrhoea. However, if the dose is too large, the arsenic will enter the organs and tissues, and the victim may develop several or all of the following symptoms: stomach pains, excessive saliva, dehydration, hoarseness and difficulty of speech, excoriation of the anus, burning pains in the urinary organs, convulsions and cramps, clammy sweats, pallor of the hands and feet, and delirium. Within one or two days, the victim falls into a coma and dies of heart failure.

KILL OR CURE Because in small amounts arsenic trioxide acts as a stimulant to the metabolism, it has found a role in medicine since antiquity. Traditional remedies from India and China have been found to contain arsenic, though usually in doses that would not cause poisoning. Arsenic is also present in certain spa waters as a trace element. In the late eighteenth century, a Dr Fowler created a tonic, known as Fowler's Solution, that contained arsenic trioxide. It remained in use until the end of the nineteenth century. It was safe in small doses but probably led to many cases of unwitting self-poisoning by people who took large quantities of the mixture thinking they were protecting their health. It is possible that one high-profile victim of the Solution was Charles Darwin (1809–82), who suffered mysterious bouts of ill health throughout his adult life. In the early twentieth century, arsenical compounds were used to treat syphilis and tropical diseases; however, the side effects of treatment could be fatal, and these drugs were phased out when antibiotics were introduced.

During the nineteenth century, travellers to the Styrian Alps in southeastern Austria reported to an incredulous world that local peasants ate large doses of arsenic trioxide by sprinkling it on their food like salt. By habituating themselves slowly over a period of years, they could tolerate very high quantities, well above the usual fatal dose, without any apparent ill effects. The Styrians took arsenic in the belief that it increased energy and potency in men, and made women plumper and gave them a better complexion.

This gave rise to the 'Styrian defence', which was used in several high-profile nineteenth-century cases of poisoning, in which the defence attorney claimed that the victim had taken the arsenic as a tonic and had not in fact been poisoned by the accused. In addition to providing a defence for murderers, the Styrian arsenic eaters also started the Victorian craze for women to eat mixtures of arsenic, vinegar and chalk to improve their complexions.

One of the most common uses for arsenical compounds during the eighteenth and nineteenth centuries was as pigments. The most widespread was 'Scheele's Green' (copper arsenite; $CuHAsO_3$), invented in 1775 by the German chemist Carl Wilhelm Scheele, which was used as a pigment for paper and wall hangings, household and artists' paints, and as a dye for cotton and linen. Amazingly, considering its toxicity, it was also used as a food colourant for sweets and beverages. In addition to being eaten in food and drink, arsenic poisoning from Scheele's Green could be caused by the inhalation of arsine gas (AsH_3), which was released from the pigment under certain conditions. Arsine has been implicated as the cause of death of no less a figure than Napoleon Bonaparte (1769–1821), who died while a prisoner of the British on the island of St Helena. When a sample of Napoleon's hair was analysed, it was found to contain large amounts of arsenic. French historians muttered darkly about a British assassination of their beloved emperor, but a more likely explanation was that Napoleon's luxury green bedroom wallpaper was coloured with Scheele's Green. St Helena has a damp climate, and it is possible that a fungus grew on the wallpaper, which metabolised the arsenic and excreted it as arsine gas, which then poisoned Napoleon.

There have been many cases of accidental or intentional arsenic poisoning. In 1858, the accidental use of arsenic trioxide in the manufacture of a batch of sweets led to the poisoning of over 200 people, 22 of whom died, in Bradford, England. In Manchester in 1900, beer contaminated with arsenic poisoned an astounding 6,000 people, and killed 70. And as late as 1932, wine containing the residue from an arsenic pesticide poisoned the 300-man crew of a French battleship.

IN ADDITION TO PROVIDING A DEFENCE FOR MURDERERS, THE STYRIAN ARSENIC EATERS ALSO STARTED THE VICTORIAN CRAZE FOR WOMEN TO EAT MIXTURES OF ARSENIC, VINEGAR AND CHALK TO IMPROVE THEIR COMPLEXIONS.

The most famous cases of arsenic poisoning, however, were intentional. The Borgias, who have earned the dubious title of the first 'criminal family' in history, used a poison called 'La Cantarella', which is now thought to have been a powder containing arsenic trioxide. Cesare Borgia (1476–1507) and his sister Lucrezia (1480–1519) murdered their enemies and relatives alike with the powder – Lucrezia dosing her victims from a hollowed-out ring. In the late seventeenth century, Giulia Toffana is reputed to have been responsible for the deaths of around 500 inconvenient husbands by selling 'Aqua Toffana' (Toffana water), a lethal cocktail of arsenic and belladonna, to would-be widows.

Arsenic's popularity endured into the modern era. One of the most famous trials of the Victorian era was that of Florence Maybrick, a beautiful American woman who was charged with the murder of her husband James, some 23 years her senior. Riven by affairs, their marriage had broken down, and Florence apparently came to the conclusion that the solution lay in the packets of arsenic poison intended for stray cats and the arsenic-based flypapers that she had bought.

© Public Domain

FLORENCE MAYBRICK
Illustrations of Florence Maybrick and her husband James, who she murdered, published in the British illustrated newspaper *The Graphic* in 1869.

Taking advantage of her husband's use of arsenical medicine as a cover, Florence proceeded to administer James a course of poison. He fell gravely ill and died after lingering for a fortnight, apparently coming to the grim realisation at the very last and uttering, 'Oh Bunny, Bunny, how could you do it? I did not think it of you.'

Florence's lawyer was also the prosecutor in the trial of Mary Ann Cotton – one of the nineteenth century's most notorious arsenic poisoners (see quote) – who murdered at least 20 people. In spite of her lawyer's experience and his use of the Styrian defence, Florence Maybrick was found guilty. She was sentenced to death, but public outcry about the judge's conduct, and the double standards demanded of men and women, meant that the sentence was commuted to life imprisonment. Arsenic's popularity as a poison declined rapidly in the nineteenth century when reliable tests to identify it were discovered.

TIPTOE THROUGH THE TULIPS: THE SCOURGE OF LANDMINES

FAILING

Never got off the drawing board

Didn't work in practice

Killed its inventor

A commercial failure

Unforeseen consequences

Was used for evil ends

A success born of failure

Main Culprits: The military

Motivation: Power and glory

Damage Done: Death and injury to soldiers and civilians alike

'Pick a place where the enemy will have to pass through, dig pits, and bury several dozen mines. All the mines are connected by fuses and originate from a steel wheel. This must be well concealed from the enemy. On triggering the firing device the mines will explode, sending pieces of iron flying in all directions and shooting flames toward the sky.'

Fire-Drake Manual, 1344, quoted in C. Ponting, *Gunpowder*

With the invention of gunpowder (see pp. 38–43), the Chinese had at their disposal an incendiary and basic explosive. As we have seen, they used it in a range of weapons, including the first proto-guns, bombs, cannons and handguns. Another deadly invention of the Chinese is the landmine, the first securely recorded use of which dates to the late thirteenth century. It took the Europeans another three centuries to come up with their own designs but, once created, they continually improved in reliability and killing power.

Modern landmines developed during World War II (1939–45), and they have been extensively used in conflicts in the twentieth century, notably in Vietnam, Cambodia and Afghanistan, where landmines dating back to wars decades old are still killing and injuring innocent civilians. Landmines can also be seen as the inspiration for the improvised explosive devices (IEDs) that have killed so many service personnel and civilians in the conflicts in Afghanistan and Iraq.

HIDDEN HORROR | The idea of using a hidden device to hamper, injure or kill your enemy was not original to the Chinese. It has been used in warfare all over the world since the most ancient times and was probably modelled on pits used by prehistoric humans to hunt and trap animals. The Romans were masters of the hidden trap. At the siege of the Gaulish city of Alesia in 52 BCE, Julius Caesar (100–44 BCE) erected a complex double ring of fortifications and siege works – one to contain the defenders and the other to hold off a relieving army. In addition to palisades, moats and ditches, the legionaries dug pits of different sizes equipped with spikes and upright stakes to impede infantry and cavalry charges. The Romans also used caltrops, a weapon consisting of two or more metal spikes welded together so that one spike would always point upward. The caltrop remained a popular weapon through the Middle Ages; it played a decisive role in the Scots victory over a mounted English army at the Battle of Bannockburn in 1314.

The great innovation achieved by the Chinese was not only to add an explosive charge to the underground trap, but also to invent ingenious ways by which enemy soldiers would trigger them. The first landmines were developed around 1250 to protect Chinese cities from the incursions of Central Asian nomads. Rather than true landmines, these were underground bombs that had to be triggered remotely by someone lighting a fuse. This made the weapon unreliable, as the fuse could either go out, the artificer might be killed before doing his job, or the enemy might not have reached the site of the mine or have passed it by the time it exploded. Chinese engineers solved these problems around 1300 with the invention of the 'invincible ground-thunder mine', the first pressure-activated landmine in history (see quote). A concealed board was the trigger that primed the firing mechanism. When an unwary soldier stepped on the board, it released weights that turned two wheels fitted with flints. If all went to plan, the flints sparked, lighting the fuses that detonated the whole minefield.

The Chinese developed other ingenious means to make their enemies trigger landmines, including booby traps. In one such device, spears were half buried in a mound. When the enemy disturbed the spears, they tipped a bowl of slow-burning material onto the fuses – and boom! Technologically the most challenging and intricate invention, given the materials technology of the day, was the first naval mine, the 'submarine dragon king', developed in the mid-fourteenth century. The mine had to be encased in an ox bladder to ensure that the powder remained dry, and floated on a weighted board to keep it low in the water. An incense stick inside a watertight piece of goat's intestine was used to light the fuse. The artificers floated the mines down river, where they would collide with enemy ships and hopefully explode.

SUBMARINE DRAGON KING

Fortunately for any passing civilians, Chinese mines suffered from the propensity of gunpowder to absorb water, which meant they quickly lost their explosive capabilities.

Despite the introduction of gunpowder technology to Western Europe in the early fourteenth century, the European military powers were slow to develop landmines. The first experimental models were made in Sicily around 1530. These were *fougasse* mines, a type of underground cannon consisting of a buried explosive charge that showered stones and gravel over a wide area, and which were particularly effective against a massed attack.

The problems with the *fougasse* were the same as with early Chinese landmines: the fuse had to be timed so that the mine exploded when the enemy troops reached the mine, the soldiers lighting the fuses could be attacked, and the mines themselves were unreliable. In 1573, three centuries after the first Chinese model, the German artificer Samuel Zimmermann designed the first fully functional self-tripping landmine in Europe, the *fladdermine* (flying mine), which combined a tripwire, a flintlock mechanism to light the fuse, and a *fougasse* filled with gunpowder and projectiles.

The breakthrough in landmine technology came in the second half of the nineteenth century, when water-susceptible gunpowder was replaced by longer-lasting high explosives, and a workable electrical firing system replaced flints and fuses. The invention of the first modern generation of landmines, known as 'torpedoes', is credited to the Confederate officer Brigadier-General Gabriel J. Rains (1803–81) in the American Civil War. In preparation for the Battle of Yorktown in 1862, Rains modified artillery shells so that they would detonate when an advancing enemy soldier tripped on a wire or stepped on the mine. On May 4, a Union cavalryman scouting along the Yorktown road became the first person to be killed by a pressure-operated landmine in the United States. The British quickly adopted the new weapon, which they used in their African campaigns in the late nineteenth century. By the turn of the twentieth century, all the major industrialised nations had added land and sea mines to their arsenals.

Landmines really came into their own during World War I (1914–18) with the development of another new weapon, the tank, which came into service with the Allies toward the end of the war. In the last year of the war, the Germans produced the first effective anti-tank mine: a wooden box measuring 18 x 14 x 8 in (46 x 35.5 x 20 cm), packed with 14 lb (6 kg) of high explosives.

GABRIEL J. RAINS
The inventor of the torpedo, a modified artillery shell that was the world's first high-explosive landmine with an electrical firing system.

The types of landmine that are in use today were developed during World War II. At the beginning of the war, both sides used anti-tank mines. However, these were cumbersome and easy to find, and could be reused by the opposing side. In order to protect them, smaller, hidden anti-personnel mines were developed. Toward the end of the war, the Allied forces advancing into Germany came across the first non-metal landmines that the Germans had laid in eastern France. In a single minefield they found 12,000 mines made out of plastic and wood, which made them difficult to find with metal detectors.

It is estimated that some 400 million mines have been manufactured since 1939, with 65 million of these deployed in conflicts during the past 20 years. World mine hotspots include Southeast Asia (Vietnam, Cambodia and Laos), Africa (Angola and Mozambique), and Western and Central Asia (Iran, Iraq, Afghanistan and Pakistan).

The tragedy of landmines is that, despite a worldwide ban, an estimated five to ten million are still being manufactured each year, while more are being deployed each day than can be cleared by de-miners.

Landmines have always inspired disgust, even among the military. During the American Civil War, General Sherman said that their use 'was not war but murder'. Improvements in technology mean that landmines are now cheaper to manufacture and last for decades. They are silent, indiscriminate killers that do not care whether the person who steps on them is a soldier, a farmer walking to his fields, or a child playing with her friends.

MINE HUNTING
An American soldier sweeps a beach with an electric landmine detector while three fellow soldiers probe the sand with their bayonets in a search for German mines in the Nottuno area of the Allied Fifth Army's beachhead in Italy, in 1943.

FAILING

Never got off the drawing board

Didn't work in practice

Killed its inventor

A commercial failure

Unforeseen consequences

Was used for evil ends

A success born of failure

ALL CHOKED UP: LEONARDO DA VINCI REINVENTS CHEMICAL WARFARE

Main Culprits: Leonardo da Vinci and the military

Motivation: Power and glory

Damage Done: Millions killed and injured

'Armis bella non venenis geri debere.' ('War should be fought with weapons, not with poisons.')

An extract from the speech given to the Roman Senate during hostilities with Germanic tribes who had used poison against Roman legionaries. From the works of Valerius Maximus, c. 30 CE

The deadly stocks in trade of chemical warfare are non-organic gaseous, liquid or solid agents and non-living organic toxins. Along with its evil twin, biological warfare (see pp. 79–84), chemical warfare must rank among history's most deplorable inventions. Its totally indiscriminate weapons kill, paralyse, blind and injure friend and foe alike, adult or child, human or animal. Although they have been subject to censures, bans and treaties since antiquity, as demonstrated by the declaration of an anonymous Roman senator quoted opposite, chemical weapons are still manufactured and used today.

Although it might seem unfair to blame the great Leonardo da Vinci (1452–1519) for its invention, as chemical warfare had been practised for millennia before his birth, he is responsible for the idea's resurrection in the early modern period with the design for a chemical weapon for naval warfare. However, it was not until the twentieth century that the full horror of full-scale chemical war was revealed during World War I (1914–18), when both sides used enormous quantities of poison gas. Chemical agents have been used in only a handful of the 200 conflicts since World War II, but they continue to be manufactured, stockpiled and used, despite repeated attempts to ban both the weapons and their use.

Ancient records attest to the use of chemical weapons by the Greeks in their perennial internecine feuds. In the seventh century BCE, the Athenians poisoned a rival city's water supply, causing not death, but chronic diarrhoea among the defenders. In the fifth century BCE, they themselves became the victims of an attempted gas attack when a besieging Spartan army burned mixtures of wood, resin and sulphur under the walls of the Athenian city of Platea to disable the defenders. On this occasion, however, the unreliability of chemical weapons was fully demonstrated when the winds changed, and it was the attackers who were choked by the noxious fumes.

FIGHTING WITH POISON

Halfway around the world, the Chinese mixed poisons, including arsenic, with black powder in order to create poisonous weapons; the Mongols brought this technology with them as they roved westward into Europe during the thirteenth century. For the most part, the relatively low-tech delivery systems and rudimentary chemistry of antiquity and the Middle Ages ensured that chemical weapons were largely ineffective – or were just as likely to poison those who were deploying them as their intended victims.

During the Renaissance, da Vinci designed the first chemical weapon of the early modern age. He proposed that projectiles containing a poisonous powder made of chalk, arsenic and verdigris (copper chloride) could be fired at enemy ships with catapults. The projectile would shower the enemy sailors with the powder, and they would die of asphyxiation. It is not known whether Leonardo's invention was ever used in warfare.

In 1812, during the Napoleonic Wars, a British naval officer named Thomas Cochrane suggested a plan to use sulphur dioxide gas against the enemy by loading ships with sulphur and coal and setting them on fire once anchored in enemy ports. To its credit, the British government turned down the plan, commenting that, 'the use of sulphur dioxide would not be in keeping with the rules of warfare'.

THE BRITISH GOVERNMENT TURNED DOWN A PLAN TO USE SULPHUR DIOXIDE GAS AGAINST THE ENEMY, COMMENTING THAT IT 'WOULD NOT BE IN KEEPING WITH THE RULES OF WARFARE'.

With the development of long-range artillery and high explosives, the military acquired a lethally effective delivery system for chemical weapons. In 1854, the Scots chemist Lyon Playfair (1818–98) designed an ordnance shell filled with poison gas, but again the British government rejected the proposal as immoral. However, Playfair's justification of his invention is instructive:

> It is considered a legitimate mode of warfare to fill shells with molten metal which scatters among the enemy, and produced the most frightful modes of death. Why a poisonous vapour which would kill men without suffering is to be considered illegitimate warfare is incomprehensible. War is destruction, and the more destructive it can be made with the least suffering the sooner will be ended that barbarous method of protecting national rights. No doubt in time chemistry will be used to lessen the suffering of combatants, and even of criminals condemned to death.

Although the Hague Convention of 1907 expressly forbade the use of poison-gas weapons in warfare, their first large-scale deployment was a mere seven years later during World War I. At first, the combatants used non-fatal tear gas to incapacitate the enemy, but on April 22, 1915, at the Second Battle of Ypres, the Germans launched a full-scale attack using chlorine gas on a 4-mile (6-km) stretch of the front line. The attack caused approximately 15,000 casualties, 5,000 of which were fatal. Although the assault broke the unprotected French and Canadian lines, the Germans failed to capitalise on their advantage to overrun Allied positions.

Chlorine is only dangerous if inhaled into the lungs, so the deployment of gas masks quickly dispelled its initial threat. However, this only served to trigger a chemical-weapons race, with the development of ever more unpleasant and dangerous chemical agents. The two main gases that were used in the latter part of the war were phosgene and the infamous mustard gas.

Phosgene acts by disrupting respiratory function: the victim suffocates. As with chlorine, a gas mask afforded sufficient protection, but because the gas was colourless and odourless, it was much more difficult for its intended victims to detect.

US MARINES
US marines standing in ranks with gas masks attached; France, 1918.

© Public Domain | Library of Congress Prints and Photographs Division

Mustard gas is a far nastier chemical agent, which tells the lie to Playfair's claim that '… a poisonous vapour … would kill men without suffering'. Mustard gas was designed not to kill but to incapacitate and injure in a particularly horrible way. The gas is an extremely powerful blistering agent, but victims exposed to the gas did not suffer immediate ill effects. Skin exposed to the brownish-yellow gas smelling strongly of mustard began to itch between 6 and 24 hours after the initial exposure. The irritation developed into large pus-filled blisters – chemical burns caused by the gas. If the victim suffered more than 50 per cent burns over his body, he or she usually died. However, even moderate exposure meant long periods of illness and convalescence, and also increased the risk of the victim contracting cancer in later life.

Countermeasures against the gas, such as the use of conventional gas masks, were ineffective, as it was absorbed through the skin. Furthermore, mustard gas persisted for days in the environment and continued to cause sickness. If the gas contaminated a soldier's clothing and equipment, then any other soldiers that he came into contact with could also become poisoned. Toward the end of the war the gas was used in high concentrations to deny the enemy access to whole areas.

In total, the combatants deployed 125,000 tons of 50 chemical agents during World War I. The estimated number of casualties from gas attacks alone was 1,297,000, some 91,200 of which were fatalities. After the war, the revulsion against chemical warfare was so great that it was outlawed by the Geneva Convention of 1925. Nevertheless, between the two world wars, poison gas was used in several conflicts, including the Spanish campaign in Morocco (1921–7) and the second Italo–Abyssinian War (1935–6).

NERVE GAS After its defeat in World War I, the German army invested heavily in chemical-warfare research. Just before the outbreak of World War II, the Germans discovered two types of nerve gas: tabun in 1937, and sarin in 1938. Although they had developed the capability to deliver these agents, the Nazis and their allies, the Italians and the Japanese, did not use chemical weapons against the Allies for fear of reprisals in kind. The Japanese, however, did use chemical weapons, including the blister agents, mustard gas and Lewisite, in China during the Sino–Japanese War (1937–45).

After the war, the Allies found shells containing tabun, sarin and soman (developed in 1944). These became the basis for the chemical arsenals that were developed by NATO and the nations of the Warsaw Pact during the Cold War, which included the British and American V-series (VE, VG, VM and VX) of nerve agents.

Death by nerve gas is not pleasant. The symptoms include profuse sweating, the filling of the bronchi with mucus, impaired vision, nausea, diarrhoea, convulsions, paralysis and respiratory failure leading to death. The gas can kill quickly through inhalation, or slowly by penetrating the body through the skin. Protection consists of a mask and chemical-resistant suit that is both expensive and cumbersome, and suits were never made available to the general population during the period when the threat of chemical war was at its highest.

The United Nations began work on chemical disarmament in 1980, and the Chemical Weapons Convention to end the manufacture, use and stockpiling of chemical agents was signed in 1993. The genie, once out of the bottle, however, can never be forced back in. Since their invention, chemical weapons have remained a permanent threat to humanity. Chemical agents have been used at least twice in the Middle East: during the Egypt–North Yemen War (1963–7) and the Iran–Iraq War (1980–8), when an estimated 100,000 Iranians fell victim to Iraqi chemical weapons, which included mustard gas and sarin. However, in the subsequent the Gulf War (1990–91), Saddam Hussein did not use his chemical arsenal against coalition forces because he feared nuclear reprisals. In the recent past, however, he had not been so cautious with the lives of his own citizens. In the last year of the Iran–Iraq War, he ordered the attack on the Iraqi-Kurdish town of Halabja with multiple chemical agents, believed to include sarin, tabun, VX and mustard gas, killing 10 per cent of the town's population of 50,000.

The chemical option is also an attractive one for terrorist groups. Compared to nuclear weapons, chemical agents are cheaper, more accessible and easier to transport with less risk of detection. The first successful use of chemical agents by a terrorist group against a civilian population was on March 20, 1995, when the Japanese religious sect Aum Shinrikyo released sarin into the Tokyo subway system, killing 12 and injuring over 5,000. After the terrorist attacks on New York's World Trade Center on September II, 2001, Al-Qaeda has threatened the United States and its allies with attacks with chemical weapons. These have yet to materialise in the West, but in Iraq insurgents have used chlorine gas in their attacks on the coalition forces and civilian population.

IN THE LAST YEAR OF THE IRAN–IRAQ WAR, SADDAM HUSSEIN ORDERED THE ATTACK ON THE IRAQI-KURDISH TOWN OF HALABJA WITH MULTIPLE CHEMICAL AGENTS, BELIEVED TO INCLUDE SARIN, TABUN, VX AND MUSTARD GAS, KILLING 10 PER CENT OF THE TOWN'S POPULATION OF 50,000.

FAILING

Never got off the drawing board

Didn't work in practice

Killed its inventor

A commercial failure

Unforeseen consequences

Was used for evil ends

A success born of failure

ROCKET MAN:
CHINA'S FIRST ASTRONAUT

Main Culprit: Wan Hoo (c. 1500)

Motivation: Scientific exploration

Damage Done: Death of the first 'astronaut'; invention of rocketry and the development of missiles capable of carrying explosives, poison gas, biological weapons and nuclear warheads

'Every gun that's made, every warship launched, every rocket fired, signifies a theft from those who hunger and are not fed, those who are cold and not clothed. This world in arms … is spending the genius of its scientists, the sweat of its labourers.'

President Dwight D. Eisenhower, 1953

© | Dreamstime

Although inventors still strive to improve human-powered flight (see pp. 10–15), there have been many who, realising that human strength alone will never suffice for sustained flight, have sought other ways of escaping gravity's embrace. However, there was no practical method of assisted flight until the development of the hot-air balloon in the late eighteenth century, and no sustained powered flight until the Wright brothers in 1903. The prospect of failure, injury or death, however, did not deter one sixteenth-century Chinese would-be astronaut, who attempted not just the first powered flight, but also literally aimed for the moon with the application of China's medieval rocket technology.

Since World War II, rocketry has also been applied with varying degrees of success to individual transportation, with rocket and jet packs. For the most part, these have proved to be expensive and impractical failures. The first manned rocket flight into space was achieved by the Soviet Union in 1961, and the Americans reached the moon in 1969; at no small cost, however, as disaster and tragedy have always dogged the space programme. In contrast, the application of rocket technology in warfare has progressed steadily from the fire arrows of medieval China to today's intercontinental ballistic missiles (ICBMs) that are capable of bringing doom to the whole planet.

The Chinese developed proto-rockets in around 1200 CE. Half a century on, rockets had a range of 500 yd (400 m), thanks to a special kind of 'flying' black powder (see pp. 38–43), and by the fourteenth century, the Chinese had developed multi-stage, multi-warhead rockets with rocket boosters very similar to modern-day missiles and space rockets. It was only a matter of time before China's expertise in rocket science would be put to use in manned flight.

© Public Domain

Around the year 1500, a Chinese official called Wan Hoo (or Wan Hu) dreamed up the idea of flying to the moon by using 47 large rockets strapped to a wicker chair. On the appointed day chosen for the first 'space flight', Wan, splendidly dressed for the occasion, sat in his chair and instructed 47 servants to light the fuses of the rockets simultaneously – and no doubt run away very fast. The fuses burned down and there was a tremendous explosion. When the smoke had cleared, Wan had vanished. In one version of the story,

WAN HOO
According to Chinese legend, Wan Hoo used 47 rockets, each lit by a separate assistant.

no traces of either the would-be astronaut or his chair were ever found, and the writer imagined that he must be living on the moon.

In 2004 the Discovery Channel's *MythBusters* attempted to recreate Wan Hoo's groundbreaking flight by using materials available in Wan's day. On their first attempt, there was a massive explosion, and after the smoke had cleared both Buster, the crash-test dummy standing in for Wan, and the chair had completely disappeared. However, on closer inspection, the remains of the badly charred Buster and chair were found scattered around the launch pad. Later attempts with modern rockets were also dismal failures, proving conclusively that poor Wan had not lifted off into the firmament but had merely been atomised by the force of the explosion.

A MODERN TAKE ON WAN HOO'S FLYING WICKER CHAIR IS THE ROCKET- OR JET PACK, THE FIRST OF WHICH WAS DEVELOPED BY THE NAZIS DURING WORLD WAR II.

A modern take on Wan Hoo's flying wicker chair is the rocket- or jet pack, the first of which was developed by the Nazis during World War II. Since then may other models have been developed, consisting of solid- or liquid-fuel rockets or jets attached to a harness – in other words, the equivalent of strapping two very hot engines full of highly explosive fuel to your back and hoping for the best. There are several models currently on the market, including the Bell and TAM rocket belts, but these devices share the same basic problems: short flying time (about 30 seconds), the high cost of both the device and the propellant, poor manoeuvrability, and last but not least, flying below parachute altitude but well above death altitude.

One device, however, that has exceeded these limitations is Swiss pilot and flight fanatic Yves Rossy's jet-powered wings, which he first flew in 2004 – although it should be noted that Rossy does not take off unaided, rather he is dropped from an aircraft at altitude. That said, he has shown his pack to be capable of gaining altitude and performing aerobatic manoeuvres. Rossy's jet pack consists of two folding carbon-fibre wings (span 8 ft/2.4 m) with four kerosene-fuelled jets under each wing. On September 26, 2008, Rossy, also known as 'Fusionman', was the first man ever to cross the English Channel with a jet pack, covering the 22-mile (35.4-km) distance in ten minutes. But with an estimated price tag of US $285,000, don't imagine it's going to replace the family sedan anytime soon.

Although the more benign effects of rocketry – sending satellites and manned spacecraft into space for the purposes of scientific inquiry – had to wait until the mid-twentieth century, its military applications have a long and murderous history. Chinese rocket technology was exported to the West via the Mongols, who used it against the Hungarians in 1241, and the Ottoman Turks, who launched them against Constantinople in their successful siege of 1453. The Indian Tipu Sultan (1750–99) used rockets to hold off the British in the late eighteenth century, and in the nineteenth century, rockets were used at the Battle of Baltimore in 1814. The missiles fired on that occasion were the origin of the phrase 'the rockets' red glare' in 'The Star-Spangled Banner'.

The greatest advances in rocketry took place in the two decades before World War II. In 1926 Robert Goddard created the first liquid-fuel rocket, greatly increasing the range and speed of missiles. The German army became interested in rocketry in the 1930s, and its research office included the brilliant young Wernher von Braun (1912–77), who later became director of the US space programme. Its most famous creations were the V1 and V2 rockets. The V2 had a range of 185 miles (300 km) and carried a 2,220 lb (1,000 kg) payload of high explosives. During the closing years of the war, V2s launched from France and Belgium rained down death and destruction on London and the southeast of England.

V2s

Today's military arsenals include a huge range of rocket designs, from small ground-to-air and air-to-air missiles to ICBMs capable of carrying huge nuclear payloads into space.

FAILING

Never got off the drawing board

Didn't work in practice

Killed its inventor

A commercial failure

Unforeseen consequences

Was used for evil ends

A success born of failure

IT'S A MAD, MAD WORLD: TULIP MANIA AND OTHER FINANCIAL FOLLIES

Main Culprits: Financiers, speculators and investors

Motivation: Greed

Damage Done: Bankruptcies on an international scale

'Men, it has been well said, think in herds; it will be seen that they go mad in herds, while they only recover their senses slowly, and one by one!'

Charles Mackay on tulip mania, *Extraordinary Popular Delusions and the Madness of Crowds*, 1841

© Erik De Graaf | Dreamstime.com

Financial instruments are nothing new; nevertheless, new ones are being invented all the time. And overly complex financial instruments that promise to make everyone rich but actually end up ruining individuals, companies and even whole economies are not new either. The first sure-fire get-rich-quick scheme to be dreamed up was also the cause of the first economic 'bubble', which was quickly followed by the first major economic 'crash'. It occurred in the Netherlands in 1636–7, and the unlikely object of the first-ever financial mania was a flower newly introduced to Europe from the Ottoman Empire, the tulip.

Establishing a pattern that has lasted until the present day, the tulip mania bubble burst and the hyper-inflated prices collapsed, causing financial ruin for many speculators and forcing the authorities to step in to regulate the market – does any of this sound familiar? Undeterred by the Dutch experience, the British and French experienced their own economic bubbles in 1720. Other notable bubbles have included the stock-market bubble of the 1920s, the Japanese asset bubble of the 1980s, the dotcom bubble of the 1990s, and the recent property bubble in the United Kingdom and United States, which is at the root of the credit crunch (see pp. 247–51).

The invention of new technologies has always been a fertile ground for the sudden inflation of asset bubbles. During the early 1840s, the United Kingdom was gripped by a speculative fever around the development of a new means of freight and passenger transport, which would go down in history as 'railway mania'. The first commercial lines to carry passengers in significant numbers were established in the 1830s, linking Britain's major industrial centres, cities and ports, and by 1840 railways were providing a good return to investors. In the next few years, banks, big business, speculators and large numbers of new middle-class investors, keen to get their slice of the economic pie, began to pour money into railway shares, without checking whether the company they were investing in was viable or even honest.

FLOWER POWER

In an interesting parallel to the present economic crisis, in 1825 the British government had deregulated the investment sector by repealing the Bubble Act of 1720, enacted after the disastrous South Sea Bubble (see below), and had failed to regulate the nascent railway industry. At the same time, the Bank of England reduced interest rates, increasing liquidity in the economy and making investment in the stock market much more attractive. Finally, the emergence of the first mass media in the form of national and local newspapers provided the ideal means for railway companies to advertise investment opportunities. Railway mania peaked in 1846, when investors were putting money into any scheme, no matter how unrealistic. When sanity prevailed, the value of railway stocks collapsed and thousands of small investors were ruined.

The tulip is a Central Asian flower first noted by Europeans in the gardens of the Sultan's Topkapi Palace in Constantinople (now Istanbul), Turkey. In the sixteenth century, merchants and diplomats brought tulip bulbs back to Europe, first to France and then to the Netherlands, where the flower became increasingly sought after as a luxury item and status symbol. In the 1590s the botanist Charles l'Ecluse established a tulip collection in the botanic gardens of the University of Leiden, creating the basis of the Dutch tulip industry. Growers quickly developed new varieties, in a dizzying array of shapes, patterns and colours, each given a grand-sounding name often prefixed with the titles 'admiral' and 'general'.

SEEDS OF FAILURE

Tulips are generally grown from bulbs (they can also be propagated from seeds, but this method produces a flower much more slowly). The flowering bulb will create a clone of itself, as well as buds, which will also develop into bulbs. The tulip has a short flowering season lasting just a week in late spring, and the new bulbs can be dug up in the summer. With the growing demand for new and rare varieties, merchants developed a range of financial innovations that are still in use in financial markets today. The spot market in actual bulbs lasted through the summer months, and for the rest of the year, merchants signed contracts for purchases of next season's bulbs – in other words, 'futures' contracts. It wasn't long before one enterprising merchant hit upon the idea of 'short selling', that is, betting that the price of a particular variety would fall rather than rise, and although the practice was banned in 1610, the regulations had to be strengthened a further three times between 1610 and 1636.

The tulip bubble began to inflate in the early 1630s, as the prices charged for the most desirable varieties rose as local and foreign speculators entered the market. By 1636 the trade was so brisk that the Dutch established the first formal futures market in Haarlem, and tulips were traded on exchanges all over the Netherlands. Prices continued to rise and then lost all contact with the real value of the traded bulbs. According to Charles Mackay (see quote), one rare bulb was traded for goods worth 2,500 florins (approximately US $35,000 at current prices), over 16 times the average annual wage for a workman. Speculative fever reached its peak over the winter of 1636–7, with people from all classes risking their savings and livelihoods on the tulip market. The lucky few made fortunes, but when the bubble burst in February 1637 and tulip prices collapsed, those left holding the bulbs lost all their money.

Humans, it seems, do not learn from past mistakes. Less than a century after tulip mania, the two major economic powers of the day, Britain and France, were both shaken by their own economic bubbles: the South Sea Bubble in the United Kingdom and the Mississippi Land Bubble in France, both in 1720. In this case, the financial instruments were stocks in limited companies. Although in both these instances, the product sold varied, the inflation of the bubble and the outcome were the same. The French scheme was the brainchild of the Scottish banker John Law (1671–1729), who created the Compagnie des Indes (Company of the Indies) in 1719. Despite its name, the company's main interests were in Louisiana, then a French colony. By vastly exaggerating the wealth of the colony, Law succeeded in triggering a speculative bonanza, which increased the share price from 500 to 15,000 livres (US $2,000 to 60,000). However, in summer 1720, confidence in the company faltered and the share price deflated to 500 livres again. Thousands lost their money and defaulted on their bank loans, shaking the French financial system to the core. The Mississippi Land Bubble is often cited as beginning the inexorable collapse of the Ancien Regime in France, which culminated in the French Revolution of 1789.

The South Sea Company was an innovative British government initiative intended to finance its growing national debt. In exchange for investors buying £10 million (US $14 million) worth of government debt, investors were given shares in the company that had been given a trade monopoly with Spanish South America. However, because Britain was more often than not at war with Spain, the trading prospects of the South Sea Company were at best poor, and at worst, non-existent.

ACCORDING TO CHARLES MACKAY, ONE RARE BULB WAS TRADED FOR GOODS WORTH 2,500 FLORINS (APPROXIMATELY US $35,000 AT CURRENT PRICES), OVER 16 TIMES THE AVERAGE ANNUAL WAGE FOR A WORKMAN.

Nevertheless, in 1720, the rumour-mill began to churn out stories of huge government investment and lucrative trading opportunities. The share price rocketed from £100 at the beginning of the year to £1,000 at its peak in August. As in the French case, the price could only go so high before reality set in. By the end of September the share price had fallen back to £150. Thousands of investors were ruined, and the bankruptcies spread to the banking sector, which had loaned money to investors and taken the company's now devalued stock as collateral.

MIGRANT MOTHER
Florence Owens Thompson, aged 32, a poverty-stricken mother of seven, and a migrant pea picker. This photograph, taken by Dorothea Lange in California in 1936, was commissioned by the Farm Security Administration and is one of the classic images of the American Great Depression.

Localised bubbles continued to occur with depressing regularity throughout the nineteenth century, but in the early twentieth century, there came the granddaddy of them all: the Wall Street Bubble of the 1920s that was punctured by the Wall Street Crash of 1929, which many economists blame for the subsequent Great Depression (lasting in the United States from 1929 to 1939). The speculative stock-market boom began in 1924. Once again, the instruments being traded were stocks, bonds and shares. By the late 1920s everyone was trying to get into the market, borrowing heavily to buy stock, which pushed prices ever higher. At its peak on September 3, 1929, the Dow Jones Average hit 381.17, registering a five-fold increase in value. The crash, when it came, was not a single traumatic event, but rather some kind of sickening roller-coaster ride, with highs and lows, rallies and falls, until the Dow Jones finally crashed off the rails, closing at its lowest point since the nineteenth century at 41.22 on July 8, 1932, having lost 89 per cent of its value.

As if to prove that we still haven't learned our lessons, the economic bubble that preceded the housing bubble of the late 2000s was the dotcom bubble of 1995–2001, which was based on the inflated valuations of Internet companies. In this case the economic instruments employed were not novel but were also stocks and shares; what was new, however, was the business model of the companies that were jockeying for investments. The dotcoms freely admitted that they did not make money and would not be able to do so for years to come, because at the beginning they would be operating at a loss to build up market share. They would recoup their losses later once they had become established and could start charging for their services. The flaw in the strategy was that many companies were aiming to dominate the same Internet sector, in which there was room for only one major player. Therefore, most of the dotcom startups were doomed to failure. The bubble burst in January 2000 when investors finally got wise to the fatal flaw in the dotcom model and rushed to get their money out.

The bursting of economic bubbles not only ruins individuals, banks and companies, it also disrupts whole economies for years. The crash that follows a bubble destroys huge amounts of wealth, which immediately impacts on consumer spending patterns. Individuals with highly valued assets spend more because they feel richer, as had been the case in the United States and United Kingdom at the height of the housing boom, but when the

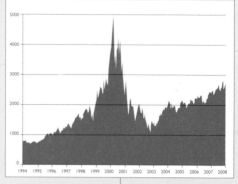

bubble bursts, those owning now depreciating assets experience a feeling of poverty and cut down on their spending, thereby hindering economic growth and worsening the economic slowdown.

Economists have put forward several explanations for our propensity to get caught up in bubbles. The first is good old-fashioned greed, which encourages people to indulge in risky financial behaviour; the second is the 'greater fool theory', which holds that there will always be someone more foolish than oneself to whom we can sell the over-priced tulip, dotcom share or house; the third is that in certain circumstances humans have a tendency to exhibit a certain lemming-like herding behaviour – and will go blithely on even if the cliff edge is clearly visible; and the fourth blames excess liquidity in the financial system, specifically lax or inappropriate lending standards by financial institutions coupled with low interest rates that make borrowing cheap and easy. To these four, one might add a fifth, that the complexity of novel financial products that not even their creators fully understand may have unexpected market effects.

THE DOTCOM BUBBLE BURSTS
A graph representing the bursting of an asset bubble: in this case the over-inflated price of dotcom stocks and their crash between 1995 and 2001. The hype that follows the invention of a new model for sure-fire financial success almost inevitably sows the seeds for its later collapse.

FAILING

Never got off the drawing board

Didn't work in practice

Killed its inventor

A commercial failure

Unforeseen consequences

Was used for evil ends

A success born of failure

COOL, FIZZING, TASTY, SWEET: COMPAGNIE DES LIMONADIERS AND FIZZY DRINKS

Main Culprits: Drinks manufacturers, fast-food outlets and lax government regulation

Motivation: Greed

Damage Done: Obesity and its related health problems: heart disease, hypertension and diabetes; cancer; hyperactivity; osteoporosis

'One extra soft drink a day gave a child a 60 per cent greater chance of becoming obese. One could even link specific amounts of soda to specific amounts of weight gain. Each daily drink added .18 points to a child's body mass index (BMI). This, the researchers noted, was regardless of what else they ate or how much they exercised. "Consumption of sugar [high-fructose corn syrup]-sweetened drinks," they concluded, "is associated with obesity in children."'

Greg Critser, *Fat Land: How Americans Became the Fattest People in the World*, 2003

© Araminta | Dreamstime.com

Itinerant vendors sold the first uncarbonated soft drink in seventeenth-century Paris. A century later, the carbonation process was invented – putting the fizz in the drink – which, along with early twentieth-century bottling technology, became the foundation for today's multi-billion-dollar fizzy-drink industry.

Carbonated drinks were once thought to have health benefits, but today's chemical products, full of contaminants, such as benzene, and additives, such as caffeine and taurine, flavourings, colourants, and natural and artificial sweeteners, have been linked to a wide range of health problems including obesity, hyperactivity in children, type-2 diabetes, tooth decay, cancer and loss of bone density. In response to the obesity epidemic in the developed world, governments have finally implemented policies to reduce fizzy-drink consumption by children, and drinks manufacturers have reluctantly responded by producing healthier alternatives and limiting their direct sales of high-calorie drinks to children.

FRENCH TOASTS

During the Middle Ages, because water supplies were often contaminated, especially in the larger towns and cities, the most common beverages for adults and children were wines in Southern Europe and beers in Northern Europe. With the gradual improvements in medical knowledge and sanitation, water became safer, and the population became more confident about drinking publicly supplied water and water-based beverages. In 1676, the French government gave a monopoly to the Compagnie des Limonadiers (company of lemonade vendors) to sell cups of *limonade* from tanks they carried on their backs to the citizens of Paris. *Limonade* was, water quality permitting, a wholesome mixture of non-carbonated water and lemon juice, sweetened with honey. The Compagnie des Limonadiers created the first mass-distribution network for non-alcoholic beverages in the early modern period, anticipating the twentieth-century's fizzy-drink vending machine and fast-food outlet.

Another early precursor of the fizzy drink was mineral water from spas, some of which were naturally carbonated. In eighteenth-century Britain, royalty visited spa towns such as Bath and Leamington Spa to 'take the waters'. The association between carbonated mineral water and health led to the first experiments in artificial carbonation, which was achieved in 1771, in the United Kingdom by Joseph Priestley, and independently by Torbern Bergman in Sweden. During the nineteenth century, artificial mineral water, which was soon christened 'soda water', quickly became popular. In the United States, flavoured soda waters were sold by pharmacists as tonics and at 'soda fountains', which became the alternative to bars for young adults, women and teetotallers. Another early soda taken for health reason was 'quinine tonic', the ancestor of modern tonic water, which was mixed with some form of alcohol, typically gin, to ward off malaria in the tropics.

© Museum of the City of New York | Corbis

ROOT BEER SELLER
Painted by Nicolino V. Calyo
c. 1840–44.

The technology to package and sell fizzy drinks as we know and love them today dates from the twentieth century. Glass bottles with sealed caps that could keep the bubbles came into common use in the 1900s; the first drinks vending machine was invented in the 1920s, and the sealed drinking can in the 1950s. With the fizzy drink available in restaurants, retail outlets, vending machines in schools, offices and public buildings, and fast-food outlets, it quickly overtook all other forms of beverage as the consumer's natural choice in the developed world. However, the switch from water, milk and juice to the fizzy drink has had some unforeseen consequences for the health of the general population.

Soft drinks contain a great deal of sugar, which equates to a lot of extra 'empty' calories in one's diet – empty because unless specially fortified, fizzy drinks contain no useful nutrients such as fibre, proteins, vitamins or antioxidants. The US Department of Agriculture (USDA) recommends the consumption of no more than ten teaspoons of refined sugar a day as part of a balanced diet, but many average fizzy-drink servings contain far more than this amount of the sweetener high-fructose corn syrup (HFCS). The danger is not just from the sugar in the fizzy drinks themselves, which can add a massive 835 extra daily calories for regular imbibers, but also because the fast-release HFCS causes an insulin spike in the bloodstream that stimulates the appetite. As fizzy drinks are often consumed with high-fat, high-sugar junk food (see pp. 150–55), this combination can cause significant weight gains in both children and adults (see quote). One report found that drinking as little as one can of fizzy drink a day equated to an increase of 1 lb (0.5kg) per month. Obesity is itself linked to the developed world's major killer, heart disease.

The negative impacts of fizzy drink on human health are not limited to obesity and heart disease. A 2004 study found that a person drinking one or more high-sugar beverages a day increased his or her chances of developing type-2 diabetes by 80 per cent. Fizzy drinks are also known to impact on dental health. The high sugar content provides the ideal nutrients for the oral bacteria that cause tooth decay, increasing the risk of cavities. In addition, many fizzy drinks are highly acidic, so the beverage itself attacks and thins the protective enamel coating of the teeth. High-caffeine fizzy drinks disrupt sleep patterns in children, and the many artificial flavourings and colourants used in fizzy drinks have been implicated in the high incidence of attention deficit disorder. It has also been suggested that acid in fizzy drink also displaces calcium in bones, leading to osteoporosis and brittle-bone syndrome. However, the evidence is not conclusive, and one researcher has argued that these conditions are observed in people who drink a lot of fizzy drinks, not because of the drink itself but because they tend to eat a diet of junk food that is very low in calcium.

> FIZZY DRINKS' HIGH SUGAR CONTENT PROVIDES THE IDEAL NUTRIENTS FOR THE ORAL BACTERIA THAT CAUSE TOOTH DECAY, INCREASING THE RISK OF CAVITIES. IN ADDITION, MANY FIZZY DRINKS ARE HIGHLY ACIDIC, SO THE BEVERAGE ITSELF ATTACKS AND THINS THE PROTECTIVE ENAMEL COATING OF THE TEETH.

Since 1990, fizzy-drink manufacturers have known of a potential cancer risk from their products caused by the carcinogen benzene (C_6H_6), which was once used as an additive in petrol. At first it was believed that benzene was present in carbonated drinks because of contaminated carbon dioxide gas, but research now suggests that it is created by chemical reactions within the beverage itself. Both the US Environmental Protection Agency (EPA) and the UK Food Standards Agency (FSA) have found products with concentrations of benzene higher than the recommended five parts per billion. The FSA banned four products in the United Kingdom, but the EPA has yet to act on its findings.

Other potential carcinogens are the by-products of quinine, a common bittering agent in fizzy drinks, which is known to break down if exposed to strong sunlight. Yet another contaminant of fizzy drinks is alcohol, which can be produced by the fermentation of sugars if the beverage is manufactured in conditions that are not completely sterile. Alcohol is also used in several of the preparatory processes for the manufacture of flavouring agents. Finally, fizzy drinks in the developing world have been found to contain dangerously high levels of pesticides.

THE BUBBLE BURSTS

Banning fizzy drinks outright, though arguably as desirable for the long-term health of humanity as the eradication of smoking, is hardly a realistic option, considering the astronomical profits made by the major food and drinks manufacturers and fast-food companies from fizzy-drink sales. Fizzy drinks are cheap to manufacture, easy to package, store and transport, so full of additives that they don't degrade, and highly addictive, so they are the ideal product of the consumer age.

Despite all the known health risks, manufacturers have fought a successful rear-guard action to keep their markets open for the past decade – in particular, their sales to children through school catering and vending machines. However, with the obesity crisis fast becoming a health catastrophe in the United States and Europe, governments have belatedly begun to act. In 2006, the United Kingdom banned the sale of fizzy drinks in school vending machines. Several US states, Canada and other major industrialised countries have made similar moves.

Sadly, it all boils down to money; legislators will act when they realise how much more inaction will cost them in healthcare costs than they receive in taxation and other financial incentives from fizzy-drink sales.

CUTTING EDGE: THE SWITCHBLADE AND GANG CULTURE

FAILING

Never got off the drawing board

Didn't work in practice

Killed its inventor

A commercial failure

Unforeseen consequences

Was used for evil ends

A success born of failure

Main Culprits: Armourers, gang members and collectors

Motivation: Power

Damage Done: Death and injury

'What an invention of the Devil is the "flick" knife, which unhappily so often features in crimes of violence in this country, often committed by young people.'

Comment made by a British judge in the 1950s, quoted in _The Times_, London, July 21, 2008

The switchblade, or flick knife, is a knife with a folding or sliding blade that opens when a button or lever is pressed on the hilt. Its distant ancestor was the eighteenth-century spring-loaded blade attached to a pistol. The stand-alone automatic knife appeared in the early nineteenth century for self-defence and military use, as well as a wide range of utilitarian functions.

In the 1940s, GIs returning from the battlefields of World War II brought back Italian stiletto switchblades, which sparked a craze for 'novelty knives' in the United States. Their popularity quickly spread to the teenage street gangs of the 1950s, which adopted the switchblade as a badge of membership. A moral panic not unlike that about the gang-and-gun culture of our own day led to the banning of switchblades in much of the developed world. Despite the ban, there is still a brisk trade in 'collectable' switchblades, which often spill out from the safety of the glass cabinet and make it onto the streets.

AN EXTRA EDGE

For centuries, personal firearms (see pp. 38–43), whether pistols or muskets, were barrel-loaded, an operation that took skill and time to accomplish. Once the single shot was fired, the gun was little better as a weapon than a wood-and-metal club, and the bearer was accordingly vulnerable. In order to give the gunner an extra edge in close combat, armourers began to add a folding spike to the barrel of pistols in the early 1700s. By the end of the eighteenth century, the combined pistol and spring-loaded blade was a standard product sold by gunsmiths in Europe and America.

In the early nineteenth century, French, English and Spanish armourers produced automatic folding knives, often with ornate blades and handles decorated with silver, ivory and mother of pearl, which opened with a variety of mechanisms. However, these were expensive one-off pieces that were beyond the reach of most purses, and bandits and cutthroats had to make do with knives and daggers that could not be concealed quite so easily. Then, in the late nineteenth century, knife production, along with the manufacture of many other consumer goods, was mechanised, bringing down their price. In the United States, automatic knives were made for farmers and tradesmen, and several smaller models were marketed to women for their sewing kits.

The Italian automatic knives brought back by servicemen returning from the battlefields of Europe in 1945 were known as *stilettos*. The *stiletto* (from the Latin *stilus*, 'spike') began life as a slender dagger, particularly favoured as a weapon during the Middle Ages and Renaissance because it could get through the chain mail and small gaps of a knight's body armour. Their modern descendants, however, were quite different creations. They had retractable blades, which, along with the handles, were richly decorated. These imported stilettos started a craze for novelty automatic knives in the United States in the late 1940s and 1950s.

The appearance of switchblades coincided with the emergence of a new type of teenage gang culture in the United States and United Kingdom, which was featured in movies such as *The Wild One* (1954), *Rebel Without a Cause* and *High School Confidential* (1955), and the Broadway musical *West Side Story* (1957). Incidents involving switchblades on both sides of the Atlantic created an atmosphere of moral panic in the popular press. As is often the case with such media-fuelled panics, the huge amount of attention given to switchblades made them even more attractive to teenagers, who adopted them as a badge of gang membership. The result was the US Switchblade Act of 1958 and the UK Offensive Weapons Act of 1959, which both prohibited the sale, and in the British case, the manufacture, of switchblades. Carrying a switchblade in public was likewise criminalised, although strangely owning an automatic knife is not illegal.

© Underwood & Underwood | Corbis

WEST SIDE STORY
A scene following a knife fight in *West Side Story* (1961).

Switchblades come in two varieties: side-opening and out-of-the-front (OTF) knives. Both are operated by a button or lever built into the hilt. The side-opening knife is hinged like an ordinary folding knife, but with the addition of a spring for quick release. In contrast, an OTF blade is stored inside the handle and slides out when the mechanism is engaged. OTF switchblades are further subdivided into single- and double-action knives. A single-action OTF must be retracted manually, and then a lever is used to compress the spring, while the double-action OTF can be extended or retracted at the press of a spring-assisted sliding button. This gives the double-action switchblade a lot more punch.

A variant on the switchblade is the 'ballistic' knife, which was developed for the Russian special forces in the 1980s. The ballistic knife can be used in several ways. Like a regular OTF knife, the blade can be extended and used as an ordinary switchblade, but the blade is also detachable and can be 'fired' from the handle by a spring-loaded or gas-powered mechanism. The ballistic knife was designed as a stealth weapon in situations when a gun could not be used. This type of weapon was added to the list of banned knives in the United States in 1986, when switchblade kits also became illegal.

BANNING BLADES

Like the current debate on gun control in the United States, arguments have raged back and forth between switchblade enthusiasts and collectors, who want the controls liberalised and bans lifted, and those who want to tighten up legislation even more. European states have by and large gone for a total ban on the manufacture, sale and carrying of switchblades, but in the United States legislation varies from one state to the next.

One could argue with the liberalisers that it is the person who wields the weapon, not the weapon itself, that is evil, but a case can also be made for there being certain weapons that are worse than others. A weapon that can easily be concealed and suddenly produced in the middle of a fight, for example, is perhaps less defensible than a weapon that is carried openly, but the distinction on moral grounds is a fine one.

The 1950s switchblade panic and subsequent bans did not bring an end to either gang violence or knife crime. The gangs in the United States moved on to firearms, and there is currently a moral panic about knife crime in the United Kingdom, where the knives that are currently in use range from illegal switchblades and military knives, to the standard kitchen knife.

POISONED CHALICE: BIOLOGICAL WARFARE

FAILING

Never got off the drawing board

Didn't work in practice

Killed its inventor

A commercial failure

Unforeseen consequences

Was used for evil ends

A success born of failure

Main culprits: The military

Motivation: Power and glory

Damage done: Death and injury to millions

'You will do well to try to inoculate the Indians by means of blankets, as well as to try every other method that can serve to extirpate this execrable race. I should be very glad if your scheme for hunting them down by dogs could take effect…'

Letter from Field-Marshal Lord Jeffrey Amherst (1717–97) to Henry Bouquet (1717–65) during the siege of Fort Pitt in 1763

© Public Domain | Centers for Disease Control and Prevention

While chemical warfare (see pp. 54–9) uses inorganic agents or inert organic toxins as weapons, biological warfare, more popularly known as germ warfare, employs living organisms, such as viruses and bacteria.

The use of biological agents in warfare began long before medical science had discovered that germs cause disease. During antiquity and the Middle Ages, combatants used human and animal corpses, venomous snakes and excrement to poison their enemies. In the New World, European settlers triggered an inadvertent biological genocide by introducing diseases to which the Native American populations had no immunity. Later, however, Europeans may well have intentionally used biological warfare against Native Americans. One instance was revealed in the correspondence between two British army officers, Lord Jeffrey Amherst and Henry Bouquet, who devised a scheme to distribute smallpox-infected blankets to Native Americans during the late eighteenth century.

Biological warfare really came into its own during World War II, when both the Japanese and German military experimented on prisoners of war and civilians to develop biological weapons. The pace of development accelerated after the war, when the United States and Soviet Union competed to create ever more lethal weapons and more efficient delivery systems. An international convention banned biological warfare in 1972, but, as with chemical warfare, the biological threat continues from rogue states and terrorist groups.

BAD AIR AND BLACK DEATH

The earliest medieval account of a biological attack in Europe dates from 1340, when England and France were fighting one of the first of the many territorial skirmishes that are now known as the Hundred Years War (1337–1453). Writing many years later, the French chronicler Jean Froissart (1337–1405) described the siege of the fortress of Thun-l'Évêque in which the attacking French side used biological weapons:

> The engines without did cast in dead horses, and beasts stinking, whereby they within had greater distress than with any other thing, for the air was hot as in the middle of summer: the stink was so abominable, that they considered how that finally they could not long endure.

The 'engines' were catapults, and doubt remains whether the French were using dead animals because they had run out of more conventional ammunition, or whether they intended to poison the enemy with the rotting corpses. The English defenders, unable to withstand the stench, abandoned the castle under cover of darkness.

The Mongols used a similar tactic at the siege of the Genoese Black Sea port of Caffa in 1346. On this occasion, however, they did not catapult dead animals into the city, but the corpses of their own men killed by an outbreak of the plague, hoping to infect the defenders with the disease.

The flea that carries the bubonic plague is thought to have originated in Central Asia, then travelled west with Mongolian merchants and armies. Historians now believe that it was the Genoese refugees fleeing from Caffa who brought the plague first to Constantinople in 1347, then to Genoa and Marseilles in 1348, triggering the most devastating epidemic of European history, the Black Death, which killed up to 60 per cent of Europe's population before burning itself out in 1351.

Similar tactics continued to be used in the succeeding centuries. In 1422, at the siege of the Bohemian town of Karlstein, the attackers sent corpses and 2,000 barrows of human excrement into the city. At the siege of the Baltic port of Reval (now the Estonian capital Tallinn) in 1710, the Russians cast plague-ridden corpses into the city. Although several of these attacks succeeded as biological assaults by infecting their intended targets, it would be over a century before the true mechanism of disease transmission would be fully understood. Contemporary theories of infectious diseases stated that they were caused by 'bad air'. Therefore the military did not have an accurate conception as to how the weapons they were using actually worked. The devastation of the Black Death is evidence enough of how indiscriminate early forms of biological warfare could be, and without inoculation against biological agents and other protective methods, the epidemic diseases often affected both sides equally.

AT THE SIEGE OF THE BALTIC PORT OF REVAL (NOW THE ESTONIAN CAPITAL TALLINN) IN 1710, THE RUSSIANS CAST PLAGUE-RIDDEN CORPSES INTO THE CITY.

The Black Death of the fourteenth century was the greatest biological holocaust the world had yet witnessed, but an even worse event was to occur just a century later. When Christopher Columbus sailed the ocean blue in 1492, he took with him much more than Christianity and trading trinkets; he carried Old World diseases to which the Native

American populations had no natural immunity. The most devastating infectious disease was smallpox, but his crew also brought typhus, measles, influenza, plague, yellow fever and whooping cough. Smallpox had a fatality rate of 20–24 per cent in fifteenth-century Europe, but without immunity or effective treatments, this figure rose to 90 per cent in the New World. In some areas, notably on the islands of the Caribbean, 95 per cent of the native population succumbed to infectious diseases, and the invading Europeans wiped out the survivors.

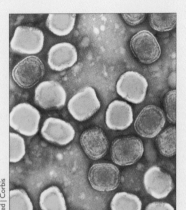

© Visuals Unlimited | Corbis

SMALLPOX VIRUSES
Photographed using a
negative-stain technique
in 1976.

The total number of fatalities in the Americas far exceeded the death toll from the Black Death, which explains why small European armies were able to overcome sophisticated Native American states with large populations in both Central and South America. This biological genocide, however, was not intentional.

In later conflicts in North America, there is stronger evidence for the intentional use of smallpox as a weapon. Although the transmission of the smallpox virus wasn't understood until the mid-nineteenth century, the correspondence between Amherst and Bouquet quoted above shows that there was some understanding of the infectious nature of the disease, and that it could be transmitted through contaminated objects, especially clothing and bedding. During the Pontiac's Rebellion of 1763, the commanding officer of Fort Pitt is said to have sent the hostile Delaware tribe blankets from a smallpox hospital just prior to an outbreak of the disease in the area. A similar accusation was made against the US Army after an outbreak of smallpox among the Mandan Indians in 1837–8. However, in neither case is the evidence of intentional infection conclusive, and the Native Americans could have come into contact with the disease when interacting with European settlers.

During the Second Sino–Japanese War and World War II with which it merged, both the Japanese and Germans experimented with biological warfare agents, despite the worldwide ban on their use that had been signed in Geneva in 1925.

Unit 731 of the Imperial Japanese Army conducted experiments on thousands of Chinese and Korean prisoners of war and civilians, and the Japanese also used bio-weapons in the field against Chinese and Russian troops.

The Nazis did not deploy biological weapons in the field, but they also conducted experiments on prisoners of war and civilians interned in concentration camps. The most infamous experimenter was Dr Josef Mengele (1911–79), who tested both chemical and biological agents on human subjects. In response to these developments, the Allies also developed bio-weapons of their own, which included the militarisation of anthrax, tularemia, brucellosis and botulism toxin.

After World War II, during the era of mutually assured destruction (MAD) known as the Cold War, both superpowers, the Soviet Union and the United States, competed to develop nuclear (see pp. 203–8), chemical and biological arsenals capable of destroying the world many times over. These programmes are still shrouded in secrecy, and the rumour, counter-rumour, propaganda and anti-propaganda issued by both sides over four decades make it difficult to sort out fact from fiction.

Did the US and UK governments test biological agents on their own servicemen and civilian populations? Did they use biological agents in Korea and Vietnam? What is known for sure is that several bio-agents were militarised, and new, even more effective delivery systems were developed.

By the 1950s, the United States had bomblets capable of dispersing the infectious diseases brucellosis, anthrax and tularemia. However, in 1972, the United States and Soviet Union signed the Biological and Toxic Weapons Convention banning the development, production and stockpiling of bio-weapons. By 1996, 137 countries had signed the convention. The current status of bio-warfare research in the major powers is unknown, but considerable funding is currently being expended on defences against bio-terrorism.

Rogue states, such as Saddam Hussein's Iraq, developed their own arsenals of biological weapons. After the First Gulf War (1990–91), Iraq admitted that it had manufactured up to 5,000 gallons (19,000 litres)

WEAPONS OF MASS INFECTION

of concentrated botulinum toxin, of which more than half had been militarised. This is approximately three times the amount of botulinum needed to kill the entire human race – while giving it a nice wrinkle-free smile, no doubt – though fortunately a method to disperse it effectively on a mass scale does not currently exist. This stock of botulinum has yet to be found, but because it degrades quickly in storage, it is unlikely to pose any persistent threat.

ANTHRAX

Terrorists have also used biological agents, although their weapon of choice so far has been anthrax. Anthrax is a bacterial disease that can be spread by inert long-lived microscopic spores that come back to life once they have entered a suitable host. Anthrax enters the body by being inhaled, eaten or through skin abrasions. The disease is not infectious, and victims are limited to those who have come into direct contact with the spores or bacillus. However, the fatality rate for pulmonary anthrax, even with immediate treatment, is close to 100 per cent. In 2001, an anthrax attack on Congress and media outlets in the United States killed five people. Fortunately for the world, although anthrax is an extremely effective bio-weapon, its manufacture and delivery demands a high level of scientific expertise and expensive equipment.

HEIGHTS OF RIDICULE: HOW THE HAIRPIECE BROUGHT DOWN A KINGDOM

FAILING

Never got off the drawing board

Didn't work in practice

Killed its inventor

A commercial failure

Unforeseen consequences

Was used for evil ends

A success born of failure

Main Culprit: Louis XIV (1638–1715)

Motivation: Vanity

Damage Done: Hastened the French Revolution; waste of money; ridicule

'In the month of April, the Duchesse de Chartres, daughter of the Duc de Penthièvre, appeared at the opera with her hair dressed in a sentimental pouf, upon which was represented her eldest son, the Duc de Beaujolais, in his nurse's arms, a parrot pecking at a cherry, a little negro boy, and the initials worked in hair from the heads of the Ducs d'Orléans, de Chartres and de Penthièvre.'

Paul Lacroix, *France in the Eighteenth Century: Its Institutions, Customs, and Costumes*, 1876

The heads of most humans, unless disadvantaged by genetic inheritance, benefit from an ample covering of their own hair. Wigs and toupées, however, have been worn since antiquity not only to mask baldness but also to indicate social status.

Wigs fell out of favour in Western Europe after the fall of the Western Roman Empire in the fifth century, but they made a comeback in the sixteenth century. The heyday of the hairpiece, however, was the eighteenth century, when a craze for ever more extravagant wigs swept the French court of Versailles. These creations, held out as exemplars of wasteful decadence by social reformers, hastened the demise of the French monarchy.

HOTHEADS

In the human mindset, a fine head of hair, status and power have close associations. For example, from the most ancient times, men have seen flowing locks as a sign of divinity and royalty. In ancient Egypt, Assyria and Persia, the king and his courtiers wore elaborate dress wigs for state occasions, while for comfort and cleanliness in the Near East's warm climate, they often cropped or shaved their own hair. Close associations between power and virility and hair made ancient Greek and Roman men with thinning hair adopt wigs and toupées. One illustrious rug wearer was no less a figure than the great Julius Caesar (100–44 BCE). With the fall of the Western Roman Empire (traditionally dated to 476 CE), wigs vanished from European history for a millennium.

The wearing of wigs resumed in the sixteenth century in the royal courts of Northern Europe. Queen Elizabeth I of England (1533–1603) wore a tightly curled red wig in the 'Roman' style to replace her own thinning white hair, and King Louis XIII of France (1601–43) wore a shoulder-length *péruque*. In the 1660s, King Charles II of England (1630–85), who had spent much of his life in exile in France, introduced a similar fashion to England once he had been restored to the English throne. As in earlier periods, the wig had originally been intended to compensate for hair loss or thinning in both men and women, and also to deal with the problem of head lice by replacing the wearer's own hair with something that could be kept clean more easily. But during the reign of Louis XIV of France ('the Sun King'; 1638–1715), the wig was taken to a whole new dimension, which within a couple of generations would help to seal the fate of the French monarchy.

During the seventeenth century, wigs were the province of men, and anyone of importance, from the king down, had to wear one. Male hairpieces were luxuriant creations of curls cascading to the shoulders or to the mid-back and chest. Made of human hair sewn onto skullcaps, they were extremely heavy, hot and uncomfortable to wear. They had other drawbacks, as this 1667 diary entry by the English writer Samuel Pepys (1633–1703) reveals:

> I did go to the Swan; and there sent for Jervas my old periwig-maker and he did bring me a periwig; but it was full of nits, so as I was troubled to see it (it being his old fault) and did send him to make it clean.

However, carrying stowaways in one's hairpiece was not the only thing that worried the seventeenth-century gentleman about town. In an earlier entry from 1665, when bubonic plague was ravaging London, Pepys confided:

> Up, and put on my coloured silk suit, very fine, and my new periwig, bought a good while since, but darst not wear it because the plague was in Westminster when I bought it. And it is a wonder what will be the fashion after the plague is done as to periwigs, for nobody will dare to buy any haire for fear of the infection that it had been cut off the heads of people dead of the plague.

Louis XIV moved the French court to his extraordinary fantasy creation, the Palace of Versailles, some 10.5 miles (17 km), from overcrowded, dirty and unruly Paris. It was there that the Sun King developed an elaborate court protocol based on fashion and conspicuous displays of wealth.

His intention had been to prevent sedition among France's senior nobles by keeping them close to him and occupied, frittering their wealth on luxuries rather than plotting to overthrow him. In this, he was only too successful. Isolated from the world, the French elite lost touch with reality during the reigns of Louis' son and grandson, Louis XV and the ill-fated Louis XVI (1754–93).

In the late eighteenth century, men's wigs shrank to a more manageable size, with the trademark curls at the sides and a pigtail at the back, and were powdered white. But with the accession of Queen Marie Antoinette (1755–93), who came to the throne with her husband Louis

NITPICKING

XVI in 1775, women's hair fashions, and briefly men's as well, became stratospheric, and did not come down to earth until quite a few noble heads had got a lot more than a buzz cut.

In the new Austrian-born queen, the fashion designers, hairdressers, jewellers and wigmakers of Paris found an appetite for excess, luxury and novelty that more than matched their wildest (and most expensive) imaginings. As hairstyles grew taller, women had to supplement what nature had provided with hairpieces and wigs, and the additions became yet another means of conspicuous display (see quote on p. 85), as they were decorated with gems, pearls and rare exotic feathers. Any event in the news became an excuse for a yet more outrageous creation. In fact, when the French frigate the *Belle-Poule* scored a famous victory against the British Royal Navy's HMS *Arethusa* in June 1778, women appeared with a model ship under full sail, all 26 of her guns blazing, atop their heads.

© Public Domain

MARIE ANTOINETTE
Marie Antoinette, painted in 1775 by Jacques-Fabien Gautier-Dagoty.

The court realised only too late that its very visible excesses had further alienated much of the French population. Despite a last-ditch attempt that saw the queen moderate her toilette, including her headgear, the French form of absolutist monarchical government was overthrown in 1789, and the royal couple went to the guillotine in 1793 (see pp. 27–31), Marie Antoinette appropriately wearing a simple mob cap.

Dress wigs, which had been associated with the excesses of the Ancien Régime, fell out of favour in the nineteenth century, never to make a reappearance. They survive only in the formal dress of judges and lawyers in the British judicial system and in parliament. Even here, there are many in modern Britain who would echo the sentiments of President Thomas Jefferson (1743–1826) when he said: 'For Heaven's sake discard the monstrous wig which makes the English judges look like rats peeping through bunches of oakum.'

Just as the dress wig was vanishing, another form of hairpiece, the toupée, was making its appearance. The function of the toupée was quite different from that of the traditional wig. While the latter was worn as a visible sign of wealth and status, and replaced or covered the owner's own natural hair, the toupée is an attempt to hide baldness by mimicking real hair. The problem is, however, that a toupée rarely fools anyone, and instead of hiding a bald spot it actually draws attention to it.

In the past decades, science has come to the rescue of balding men with topical medications, hair weaves and hair transplants. Hairstyles have also changed, and closed-cropped or even shaved heads are now quite acceptable outside the armed forces and the prison yard. As for the dress wig, it has been relegated to the entertainments industry, fancy dress and that other arena of dressing up in wig and gown, the British courts. It cannot be long before the toupée, along with the wig, is consigned to the dressing-up box of history, along with the codpiece and the crinoline (see pp. 99–103).

FAILING

Never got off the drawing board

Didn't work in practice

Killed its inventor

A commercial failure

Unforeseen consequences

Was used for evil ends

A success born of failure

ALL IN THE MIND: THE QUACK SCIENCE OF PHRENOLOGY

Main Culprit: Franz Joseph Gall (1758–1828)

Motivation: Scientific inquiry

Damage Done: Encouraged racism and prejudice

'It is not at all likely, I think, that the travelling expert ever got any villager's character quite right, but it is a safe guess that he was always wise enough to furnish his clients character charts that would compare favourably with George Washington's. It was a long time ago, and yet I think I still remember that no phrenologist ever came across a skull in our town that fell much short of the Washington standard. This general and close approach to perfection ought to have roused suspicion, perhaps, but I do not remember that it did.'

Mark Twain on phrenology, quoted in C. Neider,
The Autobiography of Mark Twain, 1959

In 1800, Franz Joseph Gall (1758–1828) invented the 'science' of phrenology (from the ancient Greek *phrenos*, meaning 'mind', and *logos*, meaning 'knowledge'), premised on the idea that different areas of the brain corresponded to distinct personality traits and abilities, which could be read by measuring the lumps and bumps on the skull. Phrenology was most influential during the Victorian period, but from its inception was viewed with suspicion by the scientific establishment, and with advances in neuroscience, psychology and psychiatry, was relegated to the status of a pseudo-science. This did not deter true believers, however, and in the early twentieth century the inventor Henry Lavery, a convinced phrenologist, patented the 'Psychograph', a machine capable of reading the human skull according to Gall's principles, which continued in use until the middle of the century. Although not as sinister as some other pseudo-scientific belief systems, phrenology misled the credulous and could also be used to justify racism.

© Public Domain

FRANZ JOSEPH GALL
Born in Baden in what is now modern-day Germany in 1758, Gall is credited with the invention of the pseudo-science of phrenology.

The ancient Greeks believed that through *physiognomy* (from the ancient Greek *physis*, 'nature' and *gnomon*, 'interpreter') a person's character, morality and abilities could be judged from his or her personal appearance, especially from the contours of the face. No less a figure than the great philosopher Aristotle (384–322 BCE) believed that it was possible to 'infer the character of persons from their features'. The theory's popularity waxed and waned over the centuries, enjoying its greatest revival in the early modern period with the work of the Swiss pastor and physiognomist, Johann Kaspar Lavater (1741–1801). Physiognomy, however, was a mixture of folk wisdom on the lines of 'his eyes are too close together, so he must be a wrong'un', with pre-scientific observation and comparisons with animal characteristics. In 1800, however, a German doctor attempt to give a scientific basis to physiognomy by creating the 'science' of phrenology.

Franz Joseph Gall was not a madman or an early New Age charlatan; he was a physician and a serious man of science. He was the first to contend that mental functions were localised in different parts of the brain, an idea that was so revolutionary at the time that it earned him the enmity of the Catholic Church and of the Imperial Austrian government, which dismissed him from his post as a university lecturer. His misfortune, however, was to be born at the dawn of the Age of Enlightenment and of the scientific age, before the scientific method of confirming one's hypotheses and theories through careful experiment and the collection of evidence had been firmly established. He initially called his theory 'cranioscopy', but it was renamed phrenology by his disciple Johann Gaspar Spurzheim (1776–1832), who popularised Gall's ideas in France and England.

Gall was essentially correct about the localised nature of specific brain functions; for example, vision is located in the visual cortex at the

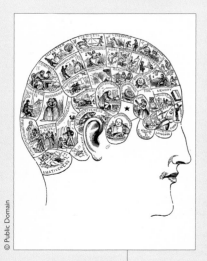

front of the brain. But in his list of 27 'brain organs', he included several for physical abilities such as speech and hearing, others for moral qualities, such as vanity, pride, benevolence and courage, and another group for such abstractions as religion, architecture and metaphysics. In this he confounded recognised brain functions with personality traits, moral qualities and fields of knowledge, which are acquired through education.

Having identified the locations of the different organs, he then jumped to the totally erroneous conclusion that their importance was reflected in the size and shape of the lumps, fissures and bumps of the skull in the immediate area. Hence, a man who was very 'benevolent' would have the corresponding area of his skull enlarged in some way.

MIND MAP
A phrenological chart from
People's Cyclopedia of Universal Knowledge (1883).

© Public Domain

The major problem with Gall's work was that it was not based on sound empirical research but entirely anecdotal evidence. He generalised from one or two personal observations, which neither he nor his disciples attempted to verify with a larger sample.

The fact that the scientific establishment immediately dismissed phrenology as a pseudo-science did not disturb Gall and his followers. During the nineteenth century it was widely used, much as psychometric testing is used today, to predict a child's scholastic ability, and to assess prospective spouses or employees. In this regard, it had just as much predictive ability as astrology or casting the runes or, some would argue, psychometric testing.

In 1901, the American Henry Lavery patented a machine that used phrenological principles to analyse an individual's character, and which he hoped would also provide the evidence that would prove its scientific validity. He completed the first of 30 machines, which he christened the 'Psychograph', in 1931. The subject sat in a chair, and the operator lowered the measuring helmet onto his or her head. Thirty-two feelers mapped the contours of the skull and fed the measurements to a recording box that did the calculations and produced an automatic printout. The box did not contain a printer, but 160 rubber stamps, one of which stamped a paper strip for each quality measured. The following was the result for a low reading for 'suavity': 'You do not make enough effort toward expressing consideration of others. Develop by studying how to please others. Exercise tact.' At the time, the machine must have made the same impression on its users as the transporter from *Star Trek* had on 1960s TV viewers – and with as much likelihood of actually working.

Phrenology might appear to be little more than a parlour game for the bored and vacuous, which is pretty much what it became in the twentieth century. But in the nineteenth, when it was taken much more seriously, it influenced decisions about people's lives. How would you feel if the size and shape of your temples affected your marriage or career prospects? Worse, proponents of eugenics (see pp. 113–17) and racism used phrenology to support their beliefs that one social class or one race was superior to another.

MISAPPLIANCE OF SCIENCE

FAILING

Never got off the drawing board

Didn't work in practice

Killed its inventor

A commercial failure

Unforeseen consequences

Was used for evil ends

A success born of failure

HOT AND BOTHERED: STEAM-POWERED CARS

Main Culprit: Nicolas-Joseph Cugnot (1725–1804)

Motivation: Scientific inquiry

Damage Done: A missed opportunity

'The mania for novelty has reached a point that is difficult to credit. A proposal has been made to substitute the horse-drawn artillery carriages with steam-powered machines, powered by pumps and pistons … This machine is as ingenious as it is useless; it is a kind of oversized cart, furnished with huge axles and wheels; the unladen weight of the contraption with its furnaces, boilers, pumps, and pistons is close to two and a half tons.'

Letter from General de St-Auban, 1779, on Cugnot's *fardier à vapeur*

With two centuries of hindsight, steam power was one the most dependable technologies developed during the Industrial Revolution. However, the early history of steam-powered transport is littered with heroic failures – who today remembers the steam-powered carriage, bus or balloon?

The ancient Greeks first understood the principle of steam power, but they considered it an interesting curiosity rather than a technological revolution that would transform their daily lives. Although primitive fixed steam turbines and engines make occasional appearances during late antiquity and the Middle Ages, the first scale model of a working steam-powered vehicle dates from seventeenth-century China, while the first full-size steam 'carriages' only appear in Europe during the late eighteenth and early nineteenth centuries. However, like many inventions that were far ahead of the technologies needed to build and operate them safely, these early designs proved impractical and unreliable and were quickly abandoned.

Steam power is an ancient technology that far predates its industrial applications in the eighteenth and nineteenth centuries. The Greek inventor and mathematician Hero of Alexandria (10–70 CE) created the first working steam turbine, the *aeolipile*, to demonstrate the principle of steam power. The first recorded application of steam to transport, however, dates to 1672, when Ferdinand Verbiest (1623–88), a Jesuit missionary working at the imperial Chinese court in Beijing, built a steam 'trolley' to amuse the 18-year-old Kangxi Emperor. The vehicle was nothing more than a scale model about 23 in (60 cm) long, and was not designed to carry a load or passengers. Most of the carrying capacity was needed to accommodate the brazier that heated the spherical boiler. The steam exited through a tube and powered a simple turbine that, in turn, drove the back wheels. As with Hero's aeolipile, Verbiest's engine was never developed. Had it been, it would have given the Chinese a century's head start on the European Industrial Revolution, and quite probably would have radically changed the course of world history.

© Public Domain

FARDIER DE CUGNOT
Cugnot designed his *fardier à vapeur* for military use, to draw ordnance to and from the battlefield.

The first practical application of steam power to transport took place in France in the late eighteenth century. As with many innovations designed in the West, it was intended for military use (see quote). The French engineer and inventor Nicolas-Joseph Cugnot (1725–1804) built the *fardier à vapeur* (a *fardier* was a heavy-duty cart used to carry or draw artillery pieces) to replace existing horse-drawn ordnance transport. He began experimenting with scale models of self-propelled steam carriages in the mid-1760s, and built the first small fardier in 1769. In 1770 he produced a full-sized version, which he hoped would travel at 4.8 mph (8 km/h) while carrying a load of 4 tons. In practice the fardier was much slower because of poor boiler performance, and its three-wheeled, front-heavy design made it difficult to steer over the poor roads of the period. One story, which may be apocryphal, relates that, in 1771, the fardier's driver lost control of his vehicle and drove it into a wall, causing what must be the first-ever 'car crash'. After several more trials, the fardier project was abandoned, the vehicles mothballed, and Cugnot pensioned off.

Cugnot's problems had been largely technological: he lacked boilers that could produce steam at sufficiently high pressures to drive his carriage at speed and for long enough periods. Fifteen years later, with the Industrial Revolution well underway in England, these problems were fast being resolved. Fixed, high-pressure steam engines were coming into use, and inventors William Murdoch (1754–1839) and Richard Trevithick (1771–1833) both designed carriages that could run on the new high-pressure 'strong steam' power. Murdoch built the first self-propelled vehicle in the British Isles, but like Verbiest's steam trolley, it was only a scale model that he never developed commercially.

© Public Domain

TREVITHICK'S STEAM CARRIAGE
Road locomotive by Richard Trevithick and Andrew Vivian, demonstrated in London in 1803.

Trevithick's first foray into steam-powered vehicles was the *Puffing Devil*, which he built in Cornwall in 1801. The vehicle weighed a hefty 1.7 tons and had a top speed of 9 mph (14.5 km/h). Although the *Devil* was technically a success, it was destroyed in a fire when it was left unattended while its driver and passengers had gone to toast the first test run in a nearby pub. Undeterred, Trevithick patented a new design the following year, using one of his strong-steam engines. The London Steam Carriage's vehicle was modelled on existing horse-drawn passenger coaches but with much larger rear wheels, some 8 ft (2.4 m) in diameter. The driver sat at the front, steering the front wheels with something that looked like a very long steering wheel, while an engineer travelled on a platform to tend the furnace and boiler mounted at the rear. The coach's inaugural run was a round trip between Paddington in west London and Islington in northeast London, carrying eight passengers at a speed of between 4 and 9 mph (6.4 and 14.4 km/h). On a subsequent run, the accident-prone Trevithick crashed his carriage into metal railings. Again, the carriage had proved itself as a workable proposition, but had failed to impress the general public. Not finding any investors, the whole scheme was abandoned, the carriage dismantled, and the engine sold off to a factory.

Accidents and commercial failures did not dishearten Trevithick. While he was working on his road steam carriages, he also built the first steam-powered locomotive for a factory in Merthyr Tydfil in South Wales in 1802. However, he must have thought that the future of transportation was on road and not on rail as he sold the patent for the locomotive in 1803. Although he built several other locomotives, Trevithick never saw their true commercial potential. The first passenger rail services would have to wait another twenty years, when George Stephenson (1781–1848) opened the Stockton and Darlington Railway in 1825.

FULL STEAM AHEAD

From the 1830s on, the commercial investment, technological research and, just as importantly, public imagination, were captivated by the railways. As we have seen above (see pp. 49–53), enthusiasm for the railways triggered an asset bubble in the United Kingdom in the 1840s. Inventors, however, did not completely abandon the idea of applying steam power to road transportation. In the 1870s and '80s, a French company designed and built several steam-powered passenger vehicles not unlike Trevithick's Steam Carriage, but with the modern arrangement of the engine at the front. When lighter and more compact steam engines came onto the market in the 1880s, several manufacturers produced three- and four-wheeled steam vehicles that held their own against early internal-combustion-driven models for a time (see pp. 108–12).

In the first two years of the twentieth century, steam-powered cars outnumbered all other types of self-propelled vehicles. There were 84 companies manufacturing 'steamers', many of which were located in New England – the 'steam Detroit' of its day. By 1903, about half of these ventures had gone out of business, unable to compete with the growing popularity of conventional automobiles. However, one US manufacturer, the Stanley Motor Carriage Company, scored a considerable success with its 'Stanley Steamers', which were manufactured from 1896 to 1924. In 1906 a Steamer broke the Land Speed Record, hitting a top speed of 127.7 mph (205.5 km/h) at Daytona Beach, Florida, a record yet to be broken by any other steam-powered road vehicle. However, with the introduction of Henry Ford's Model T in 1908, the days of steam-powered cars were clearly numbered. Mass-production methods and huge investments in technology made internal combustion vehicles ever cheaper and more efficient, while steam-power technology lagged behind.

ON THE WRONG TRACK

The steam-powered car was not so much a poor invention as one that missed its moment. At the beginning of the nineteenth century, several unfortunate accidents, as well as Trevithick's inability to market his own inventions, meant that investment and interest were diverted into the railways. The domination of rail transportation for freight and mass transit would not be challenged for more than a century. Even when advances in technology made steam cars affordable and reliable, and there was both the demand and necessary disposable income to make individual transportation a viable proposition, steam was again beaten to the finish line by another contender – internal combustion. However, it would be interesting to speculate what kind of world we would be living in today had Cugnot and Trevithick succeeded in realising their dreams for steam-powered road transportation.

SITTING PRETTY: THE CRINOLINE

FAILING

Never got off the drawing board

Didn't work in practice

Killed its inventor

A commercial failure

Unforeseen consequences

Was used for evil ends

A success born of failure

Main Culprits: Fashionistas

Motivation: Vanity

Damage Done: Ill-health; injury and death; subjugation of women

'There were also many tales of accidents that could befall hapless wearers of crinolines, such as being caught in her hoops as she descended from a carriage or of causing havoc if she were a factory girl or servant as china, glass and other delicate materials could easily be swept off shelves and tables.'

Lucy Johnstone, Curator of Fashion at the Victoria and Albert Museum, London, from 'Restrictive Flamboyance and the Crinoline Craze: 1830–1860'

The prescriptions of fashion are responsible for several of history's oddest and least functional inventions, and women have been repeatedly cajoled, forced or fooled, sometimes by men and sometimes by themselves, into being its victims. Ranking high among history's least practical forms of female attire is the nineteenth-century crinoline and its successor the bustle, both designed to dramatically transform the female silhouette, the former turning the lower portion into a large upturned fruit bowl, and the latter giving the wearer a high horse's rump. Combined with the whalebone corset, which dramatically reduced the size of the waist, wearing a crinoline meant that a Victorian woman was hampered in the most basic of movements and prone to fainting. In a very physical way, her style of dress expressed her social, political and economic status as a second-class citizen.

BASKET CASES

The ancestors of the nineteenth-century crinolines were the sixteenth-century farthingale and eighteenth-century *panniers* (from the French for 'baskets'), which were worn under a full skirt to extend it to the sides. Panniers were favoured by French aristocrats of the Ancien Régime and disappeared with them during the revolutionary and Napoleonic periods at the end of the eighteenth century and the start of the nineteenth, when simple Greco-Roman lines were preferred for women's clothing. This simple 'Regency' or 'Empire' line survived into the 1820s, by which time women had been released from the constrictions of corsetry, stiff fabrics and overly large skirts, and the discomfort of multiple heavy petticoats for 40 years. The freeing of the female body seemed to presage full political and social emancipation of women. However, such optimism was about one hundred years premature.

It is difficult to imagine why women, who had been liberated from uncomfortable, impractical garments, would once again choose to wear them. One could speculate that an increase in wealth in the early nineteenth century, combined with the emergence of the new middle class of industrial entrepreneurs, released many women from the need to work, and gave them the leisure time and income to expend on fashion. In the late 1820s, the classical androgyny of the Empire style gradually made way for a different and more highly ornamented female silhouette. Waistlines travelled downward, sleeves puffed out, and skirts swelled with petticoats and flounces to create an exaggerated feminine hourglass figure. At this critical juncture in the history of fashion, a new fabric, crinoline, made its appearance in France.

Contrary to current usage, which uses the word crinoline for the hoop armature that supports a skirt, the word was first used to describe a stiff fabric made of horsehair (the French for which is *crin*) and linen (the French for which is *lin*). Crinoline first appears in the 1830s as a material for petticoats, and the name was quickly applied to the petticoats themselves. During the 1840s and '50s skirts became fuller through the use of six or more petticoats stiffened with starch and whalebone or cane hoops and external flounces and pleats. Lacking today's light artificial fabrics, our foremothers dragged around several heavy layers of material and whalebone, which was not only weighty but must have been extremely uncomfortable in the summer months. Liberation of a sort, however, was at hand.

In 1856, the American W.S. Thomson patented the metal cage crinoline, an assemblage of metal hoops held together with fabric tape. At a stroke, women were freed from heavy petticoats, as only one was needed to smooth the outline of the metal hoops under the skirt. The cage was an instant fashion sensation, and was adopted by women from across the social spectrum. Although it considerably eased the lot of women, the cage crinoline had several drawbacks. The most obvious problem was its size. Crinolines grew to 6 ft (180 cm) in diameter, creating difficulties for going through doors, climbing in and out of carriages, and negotiating household hazards such as chairs and furniture.

Another problem was caused by the cage's lightness compared to the earlier layers of heavy petticoats. A strong gust of wind could send it flying upward, revealing that most shameful of Victorian secrets – that women had legs, even if they were encased in voluminous ankle-length bloomers. Worse, if the hapless wearer tripped and fell over, the crinoline would flip up the skirt and hold it aloft. In a similar vein, if the wearer did not spread the crinoline out properly as she sat down, the hoops would swing up to her face.

The most serious problem with crinolines, however, was that they were a major health-and-safety hazard for working women. Maids

CRINOLINES GREW TO 6 FT (180 CM) IN DIAMETER, CREATING DIFFICULTIES FOR GOING THROUGH DOORS, CLIMBING IN AND OUT OF CARRIAGES, AND NEGOTIATING HOUSEHOLD HAZARDS SUCH AS CHAIRS AND FURNITURE.

sweeping through crowded Victorian rooms might sweep away more than dust, taking ornaments and small items of furniture with them as well; pottery workers were repeatedly, and unsuccessfully, urged to leave their crinoline and hoops at home to avoid breakages; and factory workers who got their skirts caught up in machinery were sometimes maimed or killed. In an era of open-hearth fires, crinolines were also a fire hazard because the fabrics of the day – silk, muslin, satin and cotton – were highly flammable.

To wear a crinoline might have been considered a trial in itself, but they were combined with another instrument of nineteenth-century torture, the corset. The corset was stiffened with whalebone and was tightly laced from the rear to give the desired hourglass figure. The lacing became so extreme that it compressed the ribcage and internal organs and reduced lung capacity, which made women susceptible to hyperventilation and fainting – referred to at the time as a 'fit of the vapours'. The problem was so common that at fashionable events there was a special room with recliners and couches set aside for the afflicted female guests.

© Hulton-Deutsch Collection | Corbis

CRINOLINE
A late nineteenth-century illustration depicts one of the drawbacks of 1850s ladies' fashions. A gentleman calls for a gate to be opened to enable his companion, who sports a broad crinoline, to exit from a field.

The stereotype of the Victorian woman who could do little but sit and simper while men cosseted her is accurate only for the women of a certain social class, but it affected the perception of all women during the period. It would not be an exaggeration to say that through their choice of clothing, women accepted their subjugated social status as the 'weaker sex'.

The crinoline reached its full majestic Scarlett O'Hara dimensions in the early 1860s. However, around the middle of the decade, fashion moved on, and the crinoline began to shrink, making way for the 'crinolette', which was a mini cage crinoline that extended to the ground at the rear. The crinolette was itself superseded by the bustle, a more durable fashion innovation that lasted until the end of the century.

The advantage of the bustle was that it created the impression of a small waist without the drawbacks of the enormous all-round crinoline. As the century wore on, the bustle continued to grow, evolving a much more pronounced humped shape on the back of the skirt immediately below the waist, giving women a high horse-like rump that was much parodied at the time.

To the modern reader, who is probably reading this wearing jeans and a T-shirt, the idea of wearing ridiculously heavy garments, half a dead whale and a floor-length metal cage, as wide as you are tall, borders on insanity. Yet women were not forced to wear the crinoline, the bustle or the corset; they did so of their own free will after decades of comparative sartorial freedom. The crinoline still exists, though now made of lightweight plastic for the more extravagant wedding dress or ball gown. The corset, too, has survived, though appropriately for the amount of discomfort it can cause, as an item of fetish wear in the contemporary bondage and S & M world.

MODERN TRENDS

FAILING

Never got off the drawing board

Didn't work in practice

Killed its inventor

A commercial failure

Unforeseen consequences

Was used for evil ends

A success born of failure

OUT OF BREATH: THE ATMOSPHERIC RAILWAY EXPERIMENT

Main Culprits: Vermin

Motivation: Scientific inquiry

Damage Done: Missed opportunity for non-polluting transport

'The great cause of failure in the scheme was the impossibility of maintaining a vacuum in the tubes. It will scarcely be credited that the powerful engineer was baffled by enemies so contemptible as field-mice, which feasted on the tallow and ate away the leather which formed the continuous valve, so that it could not be kept air-tight. Rain, frost and sunshine also acted injuriously on the valve; and though puttymen, with pots and spatulas followed each train, the maintenance of a working vacuum was found to be impracticable. The result was that, after a loss of nearly half a million in money, the atmospheric tubes were all pulled up to give place to the locomotive engine. The failure of the scheme was a source of great grief to Mr Brunel.'

The London Quarterly Review, 1862

The pre-eminent engineer and inventor of the Age of Steam, Isambard Kingdom Brunel (1806–59), had many outstanding achievements to his name. However, his experimental version of an 'atmospheric' railway (1847–8), which used air pressure to drive trains instead of conventional locomotives, failed because it was far ahead of its time.

Although several lines using atmospheric technology were built during the nineteenth century, they were abandoned because of technical problems and high costs. Two further attempts at air-powered mass transit were made later in the century in New York and London, but again without long-lasting success. Although not a 'clean' technology because it employed conventional coal-burning steam engines as its main power source, the atmospheric system was potentially less polluting than standard steam trains; had it taken off, it could have provided cleaner alternatives to forms of mass transit based on conventional steam engines and later on the internal combustion engine (see pp. 108–12).

BRUNEL
The Victorian engineer and inventor Isambard Kingdom Brunel (1806–59) stands in front of the chains of the *Great Eastern*, the ship he designed with John Scott Russell, c. 1850.

© Hulton-Deutsch Collection | Corbis

Steam locomotives belching out clouds of black smoke and sparks from their smokestacks are so much a part of our technological folklore that we are not aware that at the dawn of the Age of Steam, inventors, engineers and politicians seriously put forward alternatives to the locomotive-driven train. Moving steam engines were not popular because of the acrid polluting smoke they produced and because people living next to railway lines were afraid that sparks from the locomotives would set fire to their fields and homes. Another drawback of the early steam locomotives was that they lacked the power to draw heavy loads up steep gradients.

One alternative technology that emerged in the 1840s was the atmospheric railway that used air pressure as its main driving force. Instead of a locomotive pulling each train on the line, the steam engines that powered the atmospheric railway were housed in fixed pumping stations built at regular intervals along the line. An airtight metal pneumatic tube laid between the rails had a groove cut along its upper surface into which slotted the piston attached to the train. The air was pumped out from the tube in front of the train by the stationary engines creating a vacuum and was readmitted behind it so that the pressure differential moved the train forward. Compared to their conventional rivals, atmospheric trains were quieter, cleaner and could climb steeper gradients. This came at a cost, however, and an atmospheric railway was almost twice as expensive as a standard steam locomotive system.

The first half of the nineteenth century was the golden age of steam railways in the United Kingdom, and while most employed the conventional arrangement of a single locomotive drawing each train, several experimental atmospheric railway lines were built between the 1840s and 1870s in the United Kingdom as well as in Ireland, France and the United States. The most ambitious was the 20-mile (32-km) stretch of line between Exeter and Newton Abbot in the west of England operated by the South Devon Railway (SDR). The SDR was part of the Great Western Railway's (GWR) line from Plymouth in Devon to Britain's westernmost county of Cornwall, which was considered an industrial marvel of the Victorian age. Brunel was GWR's head engineer, and he decided on an atmospheric system for this section of the SDR line because its gradients were too steep for conventional locomotives. He built pumping stations at 3-mile (4.8-km) intervals, which he designed to drive the trains at an impressive 70 mph (112 km/h), though in practice average speeds were around 40 mph (64 km/h).

ALTHOUGH THE TECHNOLOGY WORKED IN PRACTICE AND WAS BETTER THAN EXISTING LOCOMOTIVE-DRIVEN TRAINS ON STEEP GRADIENTS, IT WAS MORE COSTLY AND MUCH LESS RELIABLE.

Although the technology worked in practice and was better than existing locomotive-driven trains on steep gradients, it was more costly and much less reliable. The main problem, however, was not cost but materials. The pneumatic tube had to be kept airtight but the materials available at the time were not up to the job. With no plastics or rubber at his disposal, Brunel had to use leather to keep the tubes airtight. Because the line was built along the coast, the salt spray damaged the leather flaps, which had to be constantly greased with tallow to keep them supple. The tallow kept the leather in good condition, but it was attractive to vermin – rats and mice – which feasted on it and the leather. The whole system, therefore, was constantly springing leaks, which caused delays and breakdowns. The services, which had started in September 1847, were abandoned in September 1848 with major losses for investors (see quote).

In the 1860s two further attempts were made to apply the atmospheric principle to passenger transit, the first in London, the second in New York. In 1864 Thomas Rammell opened the Crystal Palace Pneumatic Railway, which ran 600 yd (550 m) between the London suburbs of Sydenham and Penge. The train carriages were fitted with large bristle collars that created an airtight seal with the tunnel walls, and a large fan powered by a steam engine sucked the trains along the tunnel. A larger version of the project planned for central London was never built.

The Beach Pneumatic Transit was another experimental atmospheric underground railway that operated along a tunnel 312 ft (95 m) long under Broadway in 1870. Although New Yorkers were initially enthusiastic about the Beach Transit, long delays and official obstructions killed the project stone dead.

The atmospheric railway was much ahead of the materials technology of the nineteenth century, in particular the materials needed to make the pneumatic tubes reliably airtight. But these problems could conceivably have been overcome had there not been much cheaper alternatives: steam locomotives and, a few decades later, gasoline-powered buses and diesel-powered trains.

© Public Domain | Library of Congress Prints and Photographs Division

AIR RAIL

Plans for a covered and elevated atmospheric railway for New York City, by Dr R.H. Gilbert, published in *Frank Leslie's Illustrated Newspaper*, March 18, 1871.

FAILING

Never got off the drawing board

Didn't work in practice

Killed its inventor

A commercial failure

Unforeseen consequences

Was used for evil ends

A success born of failure

GRIDLOCK: THE INTERNAL COMBUSTION ENGINE

Main Culprits: Automobile manufacturers; oil majors; governments; consumers

Motivation: Economic development

Damage Done: Social and environmental problems that threaten the future of the planet

'When commercial trucks are stuck in traffic, the cost of shipping goods increases, resulting in higher prices at the store. Drivers also pay with their health. The stress of driving in heavy traffic can cause high blood pressure and increased heart rates.'

Michelle Lomberg, *Avoiding Gridlock,* **2003**

As we shall see elsewhere in the pages of this book, inventions can interact with one another to make one course of technological development seem inevitable. We have seen how the development of 'black powder' (see pp. 38–43) led to the creation of a number of weapons technologies. In the same way, the internal combustion engine (ICE), when combined with petroleum technology (see pp. 21–6), quickly became one of the most influential inventions of the twentieth century. Making the private automobile practical, reliable and affordable, the internal combustion engine redesigned our cities and transformed our day-to-day lifestyles. At the same time, its enormous success has had far-reaching negative impacts on the environment, human health and society. In addition to its role in contributing to pollution and global warming, the automobile has become so ubiquitous and popular that through its sheer numbers it threatens to throttle the industrial and commercial economy that it helped to create.

ICE VERSUS ECE

The signature invention of the nineteenth century was the external combustion engine (ECE) – the steam engine – that powered the Victorian age's industry and transport, and gave shape to its built environment. The external combustion engine, despite efforts that have continued to the present day, has yet to be successfully applied to automobiles (see pp. 94–8); it was, however, with a few notable failures, such as the atmospheric railway (see pp. 104–7), applied very successfully to mass transit through the steamship lines and railway networks that dominated passenger and goods transport until the first half of the twentieth century. The signature innovation of the twentieth century, however, was the petroleum-powered internal combustion engine, which in a matter of decades after its invention began to reshape the world's cities and economies.

What gave the internal combustion engine the edge over the tried-and-tested external combustion engine? In short, the external combustion engine uses a less efficient two-stage process when compared to the internal combustion engine. In a conventional steam engine, a heat source such as a coal- or gas-powered furnace heats water in a pressurised boiler into steam that drives a piston or turbine in the engine. The internal combustion engine employs a one-stage process in which the explosive combustion of fuel (most commonly gasoline) and oxygen inside a combustion chamber drives the pistons directly. The internal combustion engine is a much more efficient engine, giving it a much better power-to-weight ratio – meaning that a smaller engine will give you a lot more bang. This explains why steam cars never took off, and why steam aeroplanes literally never got off the ground.

THE FATHER OF INTERNAL COMBUSTION

The unlikely pioneer of internal combustion was a Roman Catholic priest from Tuscany, Italy, Father Eugenio Barsanti (1821–64). In 1841, while he was lecturing in science at a local college, he hit upon the idea of using the combustion of hydrogen and air to power a simple motor. Ten years later, Barsanti met hydraulics engineer Felice Matteucci (1808–87), with whom he worked to perfect the design for a new engine. They patented the Barsanti-Matteucci internal combustion engine in London in 1854. Unfortunately, Barsanti died in 1864 just as the engine was entering commercial production. Matteucci was not a businessman and retired from the project, leaving the development of the invention to others. Barsanti and Matteucci's engine design was still too large to be mounted into an individual vehicle, and their aims were to provide the plant to power fixed factory machinery and large ships. Nevertheless, they had designed an engine that was safer, smaller, more efficient and quicker to start than the steam engines of the day.

The first proto-motorcar, however, was built a mere six years after Barsanti's death. In 1870, the Austrian inventor Siegfried Marcus (1831–98) fitted a primitive internal combustion engine onto a handcart, creating an ungainly contraption that was the world's first gasoline-powered, internal-combustion-driven vehicle. Another two decades on, and automobile pioneers Karl Benz (1844–1929) and Gottlieb Daimler (1834–1900) were manufacturing the first motorcars. When Henry Ford (1863–1947) launched the first mass-production assembly line for his Model T in 1908, the automobile industry was born. It is interesting to note that the original Model T could run on either gasoline or ethanol (grain alcohol), but the lower cost of gasoline and the advent of Prohibition (1920–33) made the use of alcohol as fuel impractical.

Today, the internal combustion engine drives all of our motorised forms of transportation, including aircraft, as jets are classed among internal combustion engines. Because they use petroleum fuels, they contribute significantly to environmental pollution by producing a number of toxic by-products, including carbon monoxide and dioxide, nitrogen oxides, formaldehyde, acetaldehyde, benzene, butadiene, sulphur oxides and sulphur dioxide, and ground-level ozone. Several of these substances have known detrimental effects on human health, while carbon dioxide (CO_2) is the major cause of global warming and climate change (see petroleum, pp. 21–6). Another by-product of internal combustion engines is noise pollution, which has become a major environmental problem in our major cities. In addition to environmental and health concerns, the dominance of the automobile in personal transportation has led to traffic congestion on an unprecedented scale in both the developed and developing world, leading some researchers to predict that without some kind of control on automobile use, we shall reach total gridlock in certain parts of the world within a decade. As can be seen from the quote above, congestion has serious consequences for individuals as well as for society as a whole.

The good news, if it can be considered as such, is that rates of car ownership in the developed world are already nearing saturation point – that is, almost everyone who wants a car owns one already. In the United States, for example, the greatest rate of increase was recorded between 1960, when 411 people per 1,000 owned a car, and 2002, when this figure soared to 812 per 1,000. In the United Kingdom, figures for the same years are 137 and 515, giving total numbers of cars in circulation in 2002 as 234 million in the United States and just over 30 million in the United Kingdom. According to the Organization for Economic Cooperation and Development, barring major change in consumption patterns, the United States and United Kingdom should hit car ownership saturation in 2030, at 314 million and 44 million cars respectively. This is still a substantial increase, but it pales into insignificance when compared to what is predicted to happen in the developing world, in particular in India and China, where car ownership rates are currently extremely low (approximately 6 cars per 1,000 people in China and 9 per 1,000 in India). With a joint population in India and China of 2.4 billion, even a minor percentage increase in car ownership will considerably bump up world totals.

© Public Domain | Library of Congress Prints and Photographs Division

HENRY FORD
Photographed c.1934, Henry Ford was the founder of the Ford Motor Company, and is credited, through the introduction of the Model T Ford, with the modern mass-production assembly line.

A WORLD WITHOUT CARS?

Car numbers were estimated at around 812 million in 2002 but they are projected to reach a mind-numbing 2.08 billion cars in 2030, with a comparable increase in associated problems in fuel supplies and environmental and social problems. Worse still, this figure is nothing near saturation point for the developing world.

For the average person in the developed world, and in particular the most heavily car-dependent cultures of North America and Europe, including the internal combustion engine on a list of 'worst inventions' must seem absurd. After all, the automobile industry has been a flagship sector driving economic and technological development since the beginning of the twentieth century. Unfortunately, it is the very success of the automobile that has created a legacy of major problems for humanity: gridlock, pollution, diminishing oil resources and climate change. Dealing with these may mean that many humans may never own an automobile and that many of those who do may have to give them up.

I can imagine that many readers – dedicated drivers in particular – might argue that without the internal combustion engine we would not enjoy the many benefits of an advanced, developed industrial economy. However, I would point out that the bulk of Europe and North America's industrialisation took place in the preceding Age of Steam, which continued well into the twentieth century. The internal combustion engine earns its place here because it has hit humanity with a double whammy. The first is the mounting problem of gridlock. Although all vehicles – gasoline, diesel, natural gas, hybrid, steam and electric – contribute to planetary gridlock, the overwhelming majority of these are still powered by internal combustion engines. The second is the issue of pollution. At the beginning of the twentieth century, internal combustion vehicles replaced the potentially far cleaner external combustion technology of steam-powered cars (see pp. 94–8), because their engines were smaller and more efficient, and therefore cheaper to build. The calculation, however, failed to factor in their true cost in terms of pollution over the past century.

HUMANS AND SUPERHUMANS: THE PSEUDO-SCIENCE OF EUGENICS

FAILING

Never got off the drawing board

Didn't work in practice

Killed its inventor

A commercial failure

Unforeseen consequences

Was used for evil ends

A success born of failure

Main Culprit: Francis Galton (1822–1911)

Motivation: Improving the human race

Damage Done: Encouraged racism; led to the genocidal theories of the Nazis and the Holocaust

'I propose to show in this book that a man's natural abilities are derived by inheritance, under exactly the same limitations as are the form and physical features of the whole organic world. Consequently, as it is easy, notwithstanding those limitations, to obtain by careful selection a permanent breed of dogs or horses gifted with peculiar powers of running, or of doing anything else, so it would be quite practicable to produce a highly gifted race of men by judicious marriages during several consecutive generations.'

Francis Galton, *Hereditary Genius*, 1869

After the publication of *On the Origin of Species by Means of Natural Selection* in 1859, in which Charles Darwin (1809–82) outlined the theory of evolution by natural selection, his cousin Francis Galton (1822–1911) devised a social version of the theory that he called 'eugenic science', a term he coined from the Greek word *eugenes* (meaning 'of good stock', from *eu* 'good' and *genos* 'birth'). Confusing inherited characteristics, such as height and hair colour, with predominantly acquired abilities, such as intelligence, Galton suggested that the same selective-breeding techniques that were used for dogs and livestock could be applied to humans in order to improve the human species. In the twentieth century, his theories were used to justify the genocidal policies of the Nazis, who eliminated anyone whom they deemed inferior to the so-called Aryan 'master race'.

SIR FRANCIS GALTON
The English scientist who studied heredity, and founder of the science of eugenics.

© Bettmann | Corbis

Although genes had not been discovered when Darwin wrote *On the Origin of Species*, he described the process of genetic mutation and selection that governs the evolution of living things. In his research for *Origin*, Darwin studied long-extinct fossil creatures, trying to understand how they were related to living species, and examined the physical differences between closely related species of animals and birds, most famously 'Darwin's finches' – a group of 13 finches he found on different islands of the Galapagos Archipelago while he was the naturalist aboard HMS *Beagle*. The key word in the previous sentence is 'physical', because Darwin's work tried to account for changes in physical and not mental characteristics.

Galton, however, applied the mechanism of natural selection to humans, specifically the mental abilities and talents of individuals, which he saw as inherited in the same way as physical characteristics. Hence, if you could 'improve' the speed of a horse or change the shape of a dog's tail by selective breeding, you should be able to do the same to the intelligence of humans, and breed men of genius. He did not stop there, however; again basing his reasoning on thoroughbred animals, he concluded that the more intelligent members of the human race were also the least fertile; hence, without intervention, humanity would be doomed to reach a level of general mediocrity. Worse still, by caring for the weak, the sick and the mentally and physically disabled, Galton believed, our societies were thwarting natural selection and encouraging humanity to degenerate.

To give him his due, Galton himself never advocated a selective breeding programme for humans. He was a promoter of 'positive' eugenics, which limited itself to encouraging marriages between people whom eugenicists considered to be genetically blessed. 'Negative' eugenics, in contrast, proposed methods, such as abortion and sterilisation, that forcefully prevented the genetically disadvantaged from reproducing.

NATURE VERSUS NURTURE

The flaw in the eugenic argument is, of course, that there is no evidence whatsoever that intelligence is primarily an inherited characteristic. Despite the best efforts of biologists since the discovery of the mechanism of genetic inheritance, no conclusive proof has yet been presented for the genetic basis of a single mental characteristic, let alone one as complex as intelligence. On the contrary, the balance of the evidence suggests that for highly complex social abilities, it is nurture and not nature that is the deciding factor. Hence, when Galton presented his 'proofs' of intelligence running in families of scientists and intellectuals such as his own, he was actually demonstrating that they had benefited from many socio-economic advantages – better health and diet as infants and children, and better educational and career opportunities in later life through their inherited position and wealth, not their inherited genes.

Absence of solid proof, however, has never prevented people believing in a theory that seems to confirm deeply held prejudices. Although Galton was well intentioned if misguided, many of the people who advocated eugenics after him had more sinister ends in mind. During the nineteenth century, racists used eugenics to justify both colonial expansion – native populations were considered to be inferior to the white colonists – and segregation of ethnic groups at home. Social conservatives argued that there was no point in trying to improve the lot of the poor, who were poor because they were less intelligent. While these examples are abhorrent, much worse was to come in the shape of the Nazi eugenics programme that lasted from 1934 to 1945.

The Nazi eugenics nightmare that was to culminate in the Holocaust began as soon as Adolf Hitler (1889–1945) came to power in 1933.

In his blueprint for the future, *Mein Kampf* ('*My Fight*'; 1925), he had already outlined his attitude to eugenics:

> He who is bodily and mentally not sound and deserving may not perpetuate this misfortune in the bodies of his children. The *völkische* [racial] state has to perform the most gigantic rearing-task here. One day, however, it will appear as a deed greater than the most victorious wars of our present bourgeois era.

A MASTER RACE In his efforts to achieve a racially pure state, Hitler enacted both positive and negative eugenics policies. At first the former were relatively benign. He encouraged 'Aryan' women to have more children with awards and financial inducements. But during the war even such 'positive' methods became corrupted. In occupied countries, 'racially valuable' women were raped by German men, and 'racially valuable' children were abducted and taken to Germany. In 1933, the German government enacted the Law for the Prevention of Hereditarily Diseased Offspring, allowing for the forced sterilisation of people with a range of mental and physical conditions. It is estimated that German doctors forcibly sterilised up to 400,000 people between 1934 and the fall of the Nazi regime ten years later.

Worse still was to come after the war had broken out. In 1939, Hitler issued a decree allowing euthanasia. His targets were mentally and physically disabled children and adults. The murderous Aktion T4 programme, which began in 1939, had 'liquidated' 70,000 men, women and children within three years, and it is estimated that by the end of the war a further 205,000 patients were murdered in Germany and occupied countries.

The figure of the 275,000 people with disabilities murdered by T4's doctors pales when compared to the mind-numbing numbers of men, women and children murdered by the Nazis in the wider phenomenon now known as the Holocaust. The victims included people from ethnic groups that the Nazis considered to be 'racially inferior', including Jews, Slavs and the Roma peoples. The 'final solution of the Jewish problem' claimed six million lives in the concentration camps of the Reich. Between a quarter and half a million Roma, around 1.8 million Poles, and two to three million Russian prisoners of war also met their deaths at the hands of the Nazis. In addition to these victims of ethnic cleansing, the regime murdered between 5,000 and 15,000 gay men and lesbians, considered to suffer from a hereditary mental disease. If all these victims are added together, we reach an estimated death toll of 11 million.

As can be imagined, eugenics, which was so closely associated with the murderous policies of the Nazi regime, fell out of favour after the war. However, negative eugenics policies, including forced sterilisations, quietly continued in the developed world until the 1970s. In liberal, democratic Sweden, for example, it is believed that up 70,000 people with mental disabilities were forcibly sterilised until the practice was stopped in 1975.

With advances in genetic science, screening for hereditary diseases and defects is now standard in the developed world. Women whose unborn children are found to have serious defects are in some cases offered terminations. The possibility of selecting embryos with certain genotypes – the creation of 'designer babies' – raises the spectre of a new and more subtle form of eugenic selection. We must trust that in the hands of a liberal-democratic regime such technology will not be abused. But who knows how far eugenic genetic selection could go in a totalitarian state, which might try to create a new 'master race'.

THE NEW EUGENICS?

FAILING

Never got off the drawing board

Didn't work in practice

Killed its inventor

A commercial failure

Unforeseen consequences

Was used for evil ends

A success born of failure

GOING OFF WITH A BANG: ALFRED NOBEL AND THE BIRTH OF HIGH EXPLOSIVES

Main Culprit: Alfred Nobel (1833–96)

Motivation: Greed

Damage Done: Increased the reliability, range and killing power of weaponry

'The clear layer is glycerin. You can mix glycerin back in when you make soap. Or you can skim the glycerin off. You can mix the glycerin with nitric acid to make nitroglycerin. You can mix nitroglycerin with sodium nitrate and sawdust to make dynamite. You can blow up bridges. You can mix nitroglycerin with more nitric acid and paraffin and make gelatin explosives. You can blow up a building, easy. With enough soap, you can blow up the whole world.'

Chuck Palahniuk, *Fight Club*, 1996

For a thousand years, the only propellant and explosive available for use in weaponry was gunpowder (see pp. 38–43). Gunpowder, however, had a very serious drawback in that it had a tendency to absorb moisture, which quickly made it unreliable or unusable, so it often had to be made in the theatre of war from its constituents. Another problem was the large amount of smoke that gunpowder generated; this made gun emplacements and musketeers clearly visible to the enemy, and if the battlefield was shrouded in a thick layer of smoke, it was impossible for the commanders standing safely on nearby hills to see how many of their men they were sending to certain death. In the nineteenth century, the military establishments of the major world powers searched for the holy grail of a moisture-proof, smokeless gunpowder. Their search was to end in the most unlikely manner and in the most unlikely of places: a kitchen table.

In 1845, the Swiss-German chemist Christian Schönbein (1799–1868) was conducting an experiment with nitric and sulphuric acids on his kitchen table, despite his wife's repeated injunctions not to do so. He spilled some of the mixture and grabbed the closest thing that was at hand, his wife's cotton apron, to clean up the mess. He guiltily hung the apron to dry over the stove, no doubt hoping that he had got away with it. A few minutes later, the apron exploded, leaving little residue and, more importantly, creating no smoke. Schönbein had just invented an entirely new kind of nitrocellulose explosive, a mixture of nitric acid and the cellulose contained in the cotton fibres of the apron. Frau Schönbein's comments when she got home are not recorded.

FRAU SCHÖNBEIN'S APRON

Nitrocellulose was smokeless and was quickly developed into a new military explosive called 'guncotton'. Unfortunately, it had the annoying and nasty habit of exploding spontaneously, and for the next 20 years it destroyed munitions factories in a way that must have gladdened the heart of many a pacifist.

Two years after Frau Schönbein's apron had mysteriously disappeared – you can't really believe that Herr Schönbein told her the whole truth – the Italian chemist Ascanio Sobrero (1812–88) of Turin University invented the second chemical compound that would play an important role in the replacement of gunpowder in military and civilian uses: nitroglycerin ($C_3H_5N_3O_9$). Like nitrocellulose, nitroglycerin is a mixture of sulphuric and nitric acids with the addition, not of cellulose, but of the sugar alcohol glycerin (glycerol; $CH_5(OH)_3$). Unfortunately, nitroglycerin was even more dangerous than nitrocellulose. If made at the wrong temperature, the mixture would produce a poisonous gas and then explode. Even when in its finished state, nitroglycerin was dangerously unstable and sensitive to shock, meaning that if it was shaken about too violently, it would explode. Sobrero was so appalled by his discovery that he held back from announcing it for a year, and even then, he advised that the substance was too dangerous to use.

Enter one Alfred Nobel (1833–96), Swedish chemist and fellow scholar of Sobrero's at Turin. Undeterred by Sobrero's dire warnings, Nobel began to experiment with nitroglycerin at his family's armaments factory in 1860. This was at some personal cost; four years later, Alfred's younger brother Emil (1843–64) was an early casualty of a nitroglycerin explosion at the Nobel factory near Stockholm, Sweden. In 1865, Nobel built a new factory in north Germany where he manufactured 'blasting oil', a mixture of nitroglycerin and gunpowder. In addition to blowing up the German factory twice, a consignment of blasting oil also destroyed the San Francisco office of the Wells Fargo Company while en route to a tunnel construction site in the Sierra Nevada. The California state government banned the transport of nitroglycerin, a move that was soon adopted by other American states and countries around the world. Nobel needed to find a way to make the new explosive stable, so that it could be stored and transported safely.

THE GOOD EARTH

In 1866–7, Nobel finally discovered a way to stabilise nitroglycerin by mixing it with kieselgur. Kieselgur is an inert chalk-like sedimentary rock with absorbent qualities, which gives it one of its most common modern uses in the filtration systems of swimming pools. He named his new high explosive 'dynamite', which was three parts nitroglycerin to one part kieselgur. As any aficionado of cartoons will know, the mixture was formed into short tubular sticks about 8 in (20 cm) long, and wrapped in greaseproof paper. The mixture kieselgur makes is much less shock-sensitive, and therefore safer to transport. The new high explosive, however, was still not completely stable. In storage, dynamite would 'weep' beads of nitroglycerin that formed pools in storage areas with predictably dangerous consequences if disturbed – boom!

Because it had a blasting strength several times greater than gunpowder, dynamite was quickly adopted for civil-engineering projects, such as building tunnels and roads, and for mining and quarrying. Imitators followed, trying to evade Nobel's patents by changing the formulae and the additives used to stabilise nitroglycerin. In 1875, Nobel himself came up with 'gelignite', a gel-like mixture of nitroglycerin, guncotton (see above) and wood pulp, which had even more explosive power than dynamite. Gelignite was the first 'plastic explosive', which soon found favour with bank robbers, freedom fighters and terrorists worldwide because it could be transported safely and shaped like putty as required.

Not surprisingly, the military interested themselves in these developments in civilian explosives. They required not only smokeless gunpowder for firearms but also high-explosive shells that could pierce the armour plating of ironclad ships. The first country to replace black powder as a propellant was France. In 1886, the chemist Paul Vieille (1854–1934) invented a type of smokeless gunpowder that could be used in firearms, which he christened *poudre B* ('powder B', taken from the French *blanche*: 'white'), to differentiate it from *poudre N* (from the French *noire*: 'black'). Although a great improvement in military terms on black powder, poudre B had a twin nitrocellulose base of collodion and guncotton. Like guncotton, it became unstable, and caused several high-profile accidents, including the destruction of two French battleships at anchor in the early 1900s.

© Public Domain

ALFRED NOBEL
(Photographed by Gösta Florman.) Appalled by the effects of his own inventions, dynamite and gelignite, Alfred Nobel established the prize that bears his name in 1895.

Alfred Nobel applied himself to the problem and patented 'ballistite' in 1887. The new military propellant incorporated Sobrero's nitroglycerin to Schönbein's nitrocellulose in the form of guncotton. Nobel sold ballistite to the Italian government, but neither the French, who were using their own poudre B, nor the British, who were working on their own version of ballistite, adopted it. Two years later, Frederick Abel (1827–1902) and James Dewar (1842–1923) developed 'cordite', thus called because it was extruded in thin cord-like shapes. Cordite's formula of nitroglycerin, nitrocellulose (guncotton) and petroleum jelly was so close to Nobel's ballistite that he tried to sue the British for patent infringement. The British won on a technicality, and cordite became the leading military propellant and high explosive of the early twentieth century. During World War I, cordite was used in rifle cartridges and artillery shells. The era of industrialised mass slaughter had arrived, with barrages of high-explosive artillery shells, and later high-explosive bombs, torpedoes and rockets.

ATONEMENT   Alfred Nobel had not invented nitrocellulose and nitroglycerin. His particular genius was to take these raw scientific inventions and turn them into several of the most destructive substances known to humankind before the development of nuclear weapons (see pp. 203–8). Granted, his high explosives did a great deal of good; by facilitating the building of roads and railroads, mining and quarrying, they encouraged economic development to the general benefit of humanity. However, the military applications of high explosives sat so heavily on his conscience that in 1895 he decided to set up the Nobel prizes for peace, science and literature.

YOU SHOULD BE SO LUCKY: MINT CAKE, CORNFLAKES AND OTHER HAPPY ACCIDENTS

FAILING

Never got off the drawing board

Didn't work in practice

Killed its inventor

A commercial failure

Unforeseen consequences

Was used for evil ends

A success born of failure

Main Culprits: Joseph Wiper (c. late 1800s), the Kellogg brothers (1852–1943 and 1860–1951) and others

Motivation: Something quite unrelated to the final invention

Damage Done: None to the inventors' bank balances

'Accident is the name of the greatest of all inventors.'

Mark Twain (1835–1910)

We may sometimes wonder at how someone came up with a great idea – the Post-it® note, for example, or the cornflake; but the truth is no one actually did. No one woke up one morning thinking they were going to invent a handy sticky paper note that you can use as a reminder or a bookmark that won't fall out, a breakfast cereal in a box or any number of the other things we now take for granted. As the author Mark Twain (1835–1910) pointed out, many inventions come about by accident. The genius at play here, and it is undoubtedly required, is to realise how an accidental discovery can be turned into a useful product. In addition to someone just tinkering with chemicals on his kitchen table and blowing up his wife's apron (see pp. 118–22), many of these serendipitous inventions came about when an inventor was trying to create something completely different. Such was the case with our first subject, a confectioner from Cumbria, England.

© Reuters | Corbis

The small town of Kendal (pop. 28,000), in the English Lake District, looks much like it did when Joseph Wiper opened his confectionery factory in the mid-nineteenth century. In 1869 he was attempting to make a mint-flavoured candy known as a 'glacier mint' – so called because it is meant to be as hard and clear as a piece of ice – when something went badly wrong with the mixture of sugar, glucose, water and peppermint oil he was cooking up. Instead of a smooth, clear paste he got a brittle, whitish substance. But the ever resourceful Wiper decided to transform adversity into opportunity and marketed the product as a new type of confectionery, which he christened 'Kendal mint cake'.

ON TOP OF THE WORLD
Sir Edmund Hillary and Sherpa Tenzing Norgay smile at camp in Thyangboche, Nepal, after their legendary ascent of Mount Everest. They ate Kendal mint cake during the climb, and also left a bar on the summit as an offering to the spirit of the mountain.

The Lake District is a mountainous region of the British Isles that is popular with hikers and mountaineers. These hardy men and women took to mint cake not only because of its distinctive tangy flavour, but also because its compactness and high glucose content made it the first energy bar.

In its 150-year history, Kendal mint cake accompanied British explorers on some of the world's greatest journeys. In 1914–17, Ernest Shackleton's (1874–1922) Trans-Arctic expedition carried Kendal mint cake to the South Pole, and in 1953, Sir Edmund Hillary (1919–2008) took it to the summit of Mount Everest in the first successful assault on the world's highest peak. According to Hillary's account, he and his companion Tenzing Norgay (1914–86) ate a mint cake while sitting on the summit, and Norgay left one as an offering to the spirit of the great mountain.

However, this is far from the only culinary failure turned success. Anyone who has cooked knows that food is a natural arena for disaster. Usually the results are inedible and end up in the rubbish bin, but just occasionally a mistake turns out to be a runaway winner. The first of two further such examples occurred in 1853, when a diner at Moon's Lake House restaurant, Saratoga Springs, New York, complained that his French fries were not to his taste, being too thick and soggy. After the customer had returned several plates of fries, the chef George Crum thought he would get his own back by slicing the potatoes so thinly that they could not be eaten with a fork. However, against all expectations, the diner was delighted, and everyone else in the restaurant wanted to try a portion. The dish became known as 'Saratoga Chips', and the story must rank as the most romantic explanation for the invention of the humble potato chip, or crisp.

COOKING UP SUCCESS

The difficult customer could have been the railroad tycoon Cornelius Vanderbilt (1794–1877) or Dr John Harvey Kellogg (1852–1943), the inventors of 'cornflakes'. Kellogg was the medical superintendent of a sanatorium in Michigan. He was a Seventh Day Adventist and strict vegetarian convinced of the virtues of whole foods, grains and nuts. One day in 1894, John's brother Will (1860–1951) left some cooked wheat to go stale. Rather than throw it away, the frugal brothers decided to roll it out into sheets of dough but instead got flakes, which they later toasted. They served their 'wheat flakes' to the patients, who found them much to their liking. From there it was a short step to making flakes with other grains, including corn. Will set up a business manufacturing and selling the brothers' toasted cornflakes in 1908, starting the company whose name is synonymous with the multi-billion-dollar breakfast cereal industry.

The chemist's laboratory has also been very productive of accidental innovation. Two famous examples come from the world of adhesives. The first occurred in 1942, when Dr Harry Coover (b. 1919) was looking for a high-grade clear plastic for gun sights. He experimented with a chemical called cyanoacrylate that instantly bonded his experimental materials together without the need for heat or pressure. Six years later, he had developed the product into the world's first 'superglue'. In 1970, an almost reverse process occurred. Spencer Silver (b. 1941) was trying to create a new type of super-strong adhesive, but what he actually made was an adhesive that was so weak that something stuck down with it could be peeled off without any effort. Again it wasn't for several years that the full potential of the new substance found the ideal application in the shape of the Post-it® note.

PERHAPS THE MOST FAMOUS ACCIDENTAL MEDICAL DISCOVERY MADE IN THE LAB WAS THE FIRST ANTIBIOTIC, PENICILLIN.

Perhaps the most famous accidental medical discovery made in the lab was the first antibiotic, penicillin. In the late 1920s the Scots chemist Alexander Fleming (1881–1955) was researching a strain of bacteria known as staphylococcus. He left cultures of the bacteria in dishes and went on holiday, only to find on his return that a fungus had contaminated one of the cultures. On closer inspection, he noticed that the fungus had killed the bacteria around it. However, Fleming did not realise the full significance of his momentous discovery, and it was only some 23 years later that further research led to the mass production of antibiotics.

Probably one of the strangest accidental scientific inventions is every mother's friend, the microwave oven. Percy Spencer (1894–1970) was working on radar at the end of the Second World War when he noticed that the microwaves emitted by a component of radar called a magnetron had melted a chocolate bar in his pocket. He rigged up an experimental oven in a metal box and first tried corn kernels, making the first instant popcorn, and then an egg, which predictably exploded, creating the instant need for someone to invent a microwave oven cleaner. The first commercial microwave, the 'Radarange', marketed in 1947, was the size of a large domestic refrigerator, weighed in excess of 750 lb (340 kg), and cost US $5,000.

The above 'worst' inventions just go to show that when you get it very, very wrong, you can actually get it very, very right.

SNAKE OILS: RADAM'S 'MICROBE KILLER' AND OTHER MAGIC POTIONS

FAILING

Never got off the drawing board

Didn't work in practice

Killed its inventor

A commercial failure

Unforeseen consequences

Was used for evil ends

A success born of failure

Main Culprit: William Radam (c. late 1800s)

Motivation: Greed

Damage Done: Prevented patients from seeking proper medical help

'The career of William Radam provides an instructive example of the unclear border between vending and science. Radam, a Texas-born gardener, read about germ theory being enunciated by Pasteur and Koch and concluded that his body teemed with germs. He would eradicate them the way a gardener would kill pests, with some sweeping poison that left the plant (or himself) unharmed.'

J.N. Hays, The Burdens of Disease, 1998

The discovery that microorganisms – bacteria and viruses – are the agents that cause infectious diseases such as typhoid, cholera, smallpox and syphilis, led to a revolution in medical science in the late nineteenth century. Germ theory opened up the way for scientifically tested treatments and a full understanding of immunisation through vaccination. It also had the unforeseen negative effect of spawning new types of patent medicines that claimed to kill microbes, which were at best completely ineffective and at worst actively poisonous. One of the most famous of these cure-alls was William Radam's 'Microbe Killer'.

HUMOURING THE PATIENT

Prior to the nineteenth century, the difference between medicines, magic potions, herbal and folk remedies, and patent medicines was largely notional. Since antiquity, doctors had believed that illness was caused by imbalances of four bodily fluids known as the 'humours': black bile, yellow bile, phlegm and blood. Treatments aimed at rebalancing the humours, the most infamous of which was bleeding by cutting or with leeches. Infectious diseases, such as cholera and plague, which periodically afflicted the overcrowded, dirty cities of Europe, were put down to *miasma* – 'bad air'. What passed for medical remedies were based on folk wisdom, guesswork, magic and the sixteenth-century 'doctrine of signatures', which attributed a curative function to plants that resembled certain parts of the body. Hence the plant toothwort, parts of which resemble teeth, was thought to be good for toothache.

The first patent medicines were so called because they had received 'letters patents'; that is, the endorsement of the British Crown – much like the right to use 'by royal appointment' on a label today. They were sometimes made by doctors, and sometimes by businessmen with little or no medical background, but as medicine at the time was little better than magic, it made little difference. One eighteenth-century example was Fowler's Solution, a tonic concocted by a certain Dr Fowler of Stafford in 1786. It was recommended as a general tonic and prescribed for a range of diseases, including malaria and syphilis. Fowler's Solution did contain one very active ingredient: potassium arsenite ($KAsO_2$), a form of the poison arsenic (see pp. 44–8). As we saw in the entry on this poison, in very small doses arsenical compounds can have a beneficial stimulant effect on the metabolism; however, if taken in large enough doses, Fowler's Solution would have caused the symptoms of arsenic poisoning. In any case, arsenic would not have been effective against the infectious diseases it claimed to cure, so the effects, if they did exist, were because of the placebo effect. In addition, potassium arsenite is a known carcinogen that causes cancers of the skin and bladder.

Another term for patent or quack medicines in the United States is 'snake oils', and it may surprise the reader that there is a genuine traditional Chinese medicine made from snake extracts for joint conditions such as arthritis and bursitis. It contains an Omega-3 oil derivative, eicosapentaenoic acid (EPA), which is known to reduce inflammation and can be absorbed through the skin. Hence, Chinese snake oil, made from extracts of the Chinese water snake (*Enhydris chinensis*), which is one of the richest known natural sources of EPA, is an effective remedy for joint problems. However, its American imitators, such as 'Stanley's Snake Oil', did not contain EPA, or indeed any snake derivative at all, but mineral oil, beef fat, red pepper, turpentine and camphor. As many of the copycat snake oils were totally ineffective while making outlandish claims as cure-alls, the term soon became synonymous with quack medicines.

EASTERN PROMISE

Patent medicines – though few if any were ever patented, because that would have forced the manufacturers to reveal their less-than-impressive list of ingredients – depended on two factors for their continuing appeal: the relative ineffectiveness of medical science and the power of advertising. The manufacturers of patent medicines advertised so heavily in the newspapers that were being established all over the United States in the nineteenth century that they were partly responsible for the creation of that country's free press. Another successful marketing method was the 'medicine show' that travelled across the country mixing snake-oil salesmanship with circus skills. During the early part of the nineteenth century, the marketing of patent medicines in the United States could be divided into the traditional and the scientific. Some of the more popular preparations owed their appeal to their claims of being based on traditional Native American remedies, but the discovery of germ theory in the 1870s gave the ever-inventive salesman an entirely new sales pitch.

One of the most successful snake-oil salesmen of the closing years of the nineteenth century was the Texan William Radam. Radam was not a doctor but a gardener. When he learned about the advances in microbiology that had revealed the role of microorganisms in disease, he reasoned: 'There is but one disease and one cause of disease no matter how varied the symptoms in different cases might be.' All he needed to do was find a substance that would attack this disease-causing agent, like weedkiller kills weeds. He wrote:

'I TREATED ALL MY PATIENTS WITH THE SAME MEDICINE, JUST AS IN MY GARDEN I WOULD TREAT ALL WEEDS ALIKE.'

I treated all my patients with the same medicine, just as in my garden I would treat all weeds alike. There are endless varieties of weeds, a very large number of which are familiar to me by name, but that would not cause me to pause about their extermination, or the method of effecting it. What matters what the scientific name of a weed might be? So long as it is a weed… So it is with disease in the human body. We are not to waste time and endanger the patient's health by trifling about special symptoms: let us remove that cause, and the person will be well.

Radam advertised his 'medicine', the 'Microbe Killer', in the following terms: 'Microbe Killer is the only known antiseptic principle that will destroy the germs of disease in the blood without injury to the tissues.' In another ad, Radam recommended: 'A wineglassful after meals and at bedtime and it will prevent and cure disease by destroying bacteria, the organic life that causes fermentation and decay of the blood, the tissues, and the vital organs.' All at the reasonable price of US $3.00 a gallon jug, or $1.99 for a 40-oz (1.3-kg) bottle. The list of diseases Radam claimed to be able to cure was impressive. The Microbe Killer's label listed: 'Cancer, syphilis, yellow fever, leprosy, smallpox, constipation, asthma, headache, neuralgias, croup, mumps, measles, whooping cough, worm, diphtheria, tonsillitis, consumption, dyspepsia, indigestion, gastritis, colds and "every disease" – applied internally or externally.'

At the height of its popularity, Microbe Killer was manufactured in 17 factories around the world. Radam answered medical criticism by attacking doctors: 'Diagnosing disease is simply blindfolding the public, but physicians dare not acknowledge it, for if they did, their glorious work would be undone, their service would not be needed.' Finally, in 1912, the Federal government ordered an analysis of Microbe Killer. It was found to contain 99.38 per cent water, traces of sulphuric and hydrochloric acids, and a dash of red wine for taste and colour.

PALACES OF THE SKIES: THE TRAGIC FAILURE OF THE ZEPPELIN

FAILING

Never got off the drawing board

Didn't work in practice

Killed its inventor

A commercial failure

Unforeseen consequences

Was used for evil ends

A success born of failure

Main Culprit: Ferdinand von Zeppelin (1838–1917)

Motivation: Mass air transport

Damage Done: End of the commercial airship flights, which allowed the much noisier, dirtier and more gas-hungry jets to monopolise the skies

'It's practically standing still now. They've dropped ropes out of the nose of the ship, and it's been taken a hold of down on the field by a number of men … The back motors of the ship are just holding it, just enough to keep it from – It's burst into flames! … It's fire and it's crashing! It's crashing terrible! Oh, my! Get out of the way, please! It's burning, bursting into flames and is falling on the mooring mast, and all the folks agree that this is terrible. This is the worst of the worst catastrophes in the world! There's smoke, and there's flames, now, and the frame is crashing to the ground, not quite to the mooring mast. Oh, the humanity, and all the passengers screaming around here.'

From Herbert Morrison's radio commentary of the Hindenburg disaster in 1937

The concept of a lighter-than-air airship sailing majestically over the clouds carrying passengers and goods was a truly inspired one, and for a time, in the 1920s and '30s, it seemed that the airways would belong not to conventional prop-powered aircraft but the huge cigar-shaped rigid airships, the most famous of which were the German-built Zeppelins. Zeppelins, however, had several technological flaws, which in the long run proved fatal. After several highly publicised accidents in the late '30s, the great airships were abandoned or mothballed, and the advent of the jetliner after World War II sealed their fate. As with several might-have-beens in this book, the safe and effective realisation of the Zeppelin concept was beyond the technology of the day. However, had it been perfected, it could have changed the course of transportation history.

A LOT OF HOT AIR

As we saw in 'Flights of Fancy' (see pp. 10–15), humans have always dreamed of taking to the skies. At first they experimented with flapping and fixed wings, parachutes and gliders, but these could not achieve sustained flight or, in many cases, any form of flight at all, unless a vertical descent at terminal velocity can be characterised as 'flight'. Without an artificial power source, it was impossible to lift a heavier-than-air craft off the ground. One obvious solution was to create a lighter-than-air craft. Balloon technology is an ancient form, and the Chinese, again at the forefront of early innovation, have made floating hot-air paper lanterns since the third century CE.

Traditionally the credit for the invention of the first balloon to carry passengers is given to the French Montgolfier brothers, Joseph-Michel (1740–1810) and Jacques-Étienne (1745–99), whose hot-air 'Montgolfière' lifted off from the Palace of Versailles, near Paris, in 1783. The following year, another French balloonist, Jean-Pierre Blanchard (1753–1809), demonstrated the first hydrogen-filled balloon. After many flights in France, he successfully crossed the English Channel in 1785. Blanchard also has the distinction of being the first man to fly in a balloon in North America, in 1793. The final French pioneer of lighter-than-air flight was Henri Giffard (1825–82), who invented a blimp-shaped, hydrogen-filled dirigible powered by a steam engine. In 1852 he made the first powered, controlled flight travelling from Paris to the village of Trappes, some 17 miles (27 km) southwest of the French capital. The engine lacked power and could not make it back to Paris against a fairly weak head wind, but the powered airship principle had been proven to work.

Although the French had pioneered ballooning in the eighteenth and nineteenth centuries, it was the Germans who made it their own in the twentieth. In 1874 Count Ferdinand von Zeppelin (1838–1917) outlined a design for a rigid airship, which he began building in 1899. Unlike French dirigibles, the Zeppelins had a rigid metal armature filled with several hydrogen gas cells and covered with a protective outer skin. This arrangement meant that the craft could be much larger and could lift heavier loads. The Zeppelins were powered and steered by the then-new types of internal combustion engine (see pp. 108–12), fitted USS *Enterprise*-style on nacelles suspended from the frame. Passengers, crew and goods were carried in a compartment built onto the underside of the Zeppelin that was accessed by lowering the craft to the ground. Airship technology, though reliable, was accident-prone and cumbersome to operate. Many of the early models were lost in crashes, storms and fires, but Zeppelin persevered.

ALTHOUGH THE FRENCH HAD PIONEERED BALLOONING IN THE EIGHTEENTH AND NINETEENTH CENTURIES, IT WAS THE GERMANS WHO MADE IT THEIR OWN IN THE TWENTIETH.

Airships played a role during World War I (1914–18) as reconnaissance craft and bombers, but they were vulnerable to attack by fighter aircraft using incendiary ammunition. Although Zeppelins had caused little actual damage during the war, their psychological impact was such that the Allies included them in the postwar settlement with Germany, confiscating existing craft as part of war reparations and temporarily halting airship production and development.

With Germany bankrupt after the war, it seemed that the Zeppelins might never fly again, but in 1924 the United States saved the company with an order for a commercial airship. Zeppelin's new boss, Hugo Eckener (1868–1954), flew the *LZ 126* (later renamed the USS *Los Angeles*) from Germany to Lakehurst, New Jersey, and was greeted with a hero's ticker-tape parade in New York after the crossing. With the future of airships assured, Eckener ordered the construction of the largest airship yet built, the 776-ft (226-m) *LZ 127 Graf Zeppelin*. A master marketeer, Eckener embarked on a series of daring journeys to popularise airship travel. In 1929, he flew around the globe from Lakehurst, via Germany, Tokyo and Los Angeles, taking a stately 21 days, and in 1931, the *Graf Zeppelin* captured the popular imagination again by flying to the North Pole. Despite competition from propeller aeroplanes, Zeppelin passenger services between Europe and the Americas began in 1930, for which Eckener ordered the construction of the even larger *LZ 129 Hindenburg* in 1931. The new airship, a veritable liner of the skies, was launched in 1936.

The *Hindenburg* was an astounding 803 ft (245 m) long, just shy of the 882 ft (269 m) length of the largest ocean liner ever built, the equally ill-fated RMS *Titanic*. It had a diameter of 130 ft (41 m), and was powered by four 1,200 HP diesel engines. It was crewed by 40 to 60 hands, and could carry up to 72 passengers. In the original design, Eckener replaced the inflammable hydrogen, which had caused so many earlier disasters, with the inert gas helium. Unfortunately for Eckener, the Nazi regime had come to power in 1933, and the United States, then the only producer of helium, had embargoed supplies of the gas. He had to fall back on the tried-and-tested but highly explosive hydrogen to fill the ship's 16 huge gas cells.

The passenger deck featured individual cabins, a lounge equipped with a specially made aluminium baby grand piano, a reading room, a smoking room, a promenade deck, and a dining room, both of which offered spectacular views of the earth below. The passenger cabins were small at 6 ft 5 in x 5 ft 5 in (2 x 1.7 m) when compared to shipboard accommodation, but with two-bunk berths and a folding washbasin and table, they probably compared favourably with the sleeper compartments of the trains of the period.

THE END OF AN ERA

On May 6, 1937, the *Hindenburg* was manoeuvring to land at Lakehurst airfield during an electrical storm. Without any apparent cause, the tail of the giant airship burst into flames (see quote) and the entire structure was alight and crashing earthward in seconds. The destruction of the *Hindenburg* took 34 seconds and killed 35 of the 97 passengers and crew. The inquiry into the disaster never definitively established the cause. One line of enquiry suggested sabotage by anti-Nazi sympathisers among the passengers or crew, but the more likely explanation was the tragic interaction of the weather, hydrogen gas, and the metallic framework of the airship. The most probable scenario was that an electrical discharge in the metal framework had ignited hydrogen leaking from gas cells damaged during the landing. Another theory blamed the outer coating of the hull, which was a highly flammable mix of cellulose butyrate acetate and powdered aluminium – a compound also used as solid rocket fuel.

What the *Hindenburg* disaster had started World War II completed. The Nazis were not interested in Zeppelin technology other than for its propaganda value. The huge airships were much too vulnerable to attack to be used as military assets, and the Luftwaffe under Hermann Göring (1893–1946) concentrated its efforts on conventional fighters and bombers. After the outbreak of the war, the Zeppelins were dismantled, and their metal frames recycled for the German war effort.

GOING UP IN FLAMES
The airship *Hindenburg* explodes as it comes in to land at Lakehurst, New Jersey on May 6, 1937.

Since the war, there have been several attempts to revive the airship to transport goods and passengers, but until now the cheapness of aviation fuel has meant that the much faster jets have always dominated the skies. But in the new low-carbon future, and with soaring oil prices, our skies may once more become populated with these majestic liners of the air.

© Bettmann | Corbis

FAILING

Never got off the drawing board

Didn't work in practice

Killed its inventor

A commercial failure

Unforeseen consequences

Was used for evil ends

A success born of failure

THE WONDER DRUG GONE WRONG: HEROIN

Main Culprits: Charles R. Alder Wright (1844–94) and Felix Hoffmann (1868–1946)

Motivation: Medical research

Damage Done: Cause of the worst drug crisis in the developed world; encouraged organised crime, war and terrorism; devastation of several countries in southeast and central Asia

'Doses [of diacetylmorphine, aka, heroin] were subcutaneously injected into young dogs and rabbits … with the following general results … great prostration, fear, and sleepiness speedily following the administration, the eyes being sensitive, and pupils constricted, considerable salivation being produced in dogs, and slight tendency to vomiting in some cases, but no actual emesis. Respiration was at first quickened, but subsequently reduced, and the heart's action was diminished, and rendered irregular.'

F.M. Pierce, report on the effects of the first experimental trials of heroin on animals, 1874

The opium poppy (*Papaver somniferum*; the 'sleep-inducing poppy') is an attractive ornamental flowering plant that is also widely grown as a food crop. Its uses as a medicine, painkiller and narcotic have been known since prehistory, but it took modern scientific know-how to transform an already dangerous and highly addictive substance – raw opium – into the lethal drug heroin. Heroin is one of the most addictive psychoactive chemicals known to humans, and its effects on individuals and society have been profound – not only in terms of the costs of addiction in the developed world but also as a direct or indirect cause of wars and revolutions in Asia.

Opium has a long and tragic history as an illicit narcotic. The latex (sap) of the poppy is collected to make opium paste, which can be eaten or smoked neat or mixed with tobacco. The recreational taking of opium in China began among the nobility in the fifteenth century. By the eighteenth century, it had become so widespread among all social groups that it was the subject of the world's first drugs ban on importation and sale (though not consumption), enacted by the Chinese authorities in 1729. The prohibition had little effect, however; not only did opium smoking increase in China but it also travelled overseas with Chinese traders and labourers.

By the nineteenth century, establishments for opium users known as 'opium dens' existed outside China in the major cities of Asia, the United States and Europe. Despite the Chinese ban, the highly lucrative importation of opium from India and Turkey continued, organised by British, American and Portuguese traders. Disputes over the trade led to two 'Opium Wars' between China and Britain (1839–42 and 1856–60). The British won, forcing the Chinese to legalise the opium trade in an act of state-sponsored terrorism that was to have far-reaching consequences for the world. The result was an epidemic of opium addiction in China that affected almost a third of the male population. Opium remained a major social problem in China until the Communist takeover of 1949, when the practice was savagely repressed with the execution of thousands of producers and dealers, and the forced treatment of addicts.

In the Western world, the smoking of opium did not take hold among the native populations. Here the most common form of opium was

OPIUM WARS

laudanum, an alcoholic tincture of opium that was highly addictive. It was sold widely as a patent medicine (see pp. 127–30), and, until it too became a controlled substance, laudanum was consumed widely as a cheap alternative to alcohol. The third narcotic to be isolated from the opium poppy in 1804 was morphine (named after the Greco-Roman god of sleep and dreams, Morpheus). At first morphine was prescribed as a treatment for opium addiction and alcoholism, until it was discovered that it was even more addictive than either opium or alcohol. Morphine quickly became the third derivative of the opium poppy to cause a major addiction problem worldwide, though it continues to be used as a painkiller.

Morphine, however, was not the last of the poppy's gifts to humanity. In 1874, the English chemist Charles R. Alder Wright was the first to synthesise a drug he called 'diacetylmorphine' (see quote). However, he did not exploit his discovery, and it was only in 1897, when Felix Hoffmann (1868–1946), a chemist working for the German pharmaceuticals company Bayer, repeated Wright's work that the drug was rediscovered and commercialised. The company called the compound 'heroin' because, according to one story, of the 'heroic' way patients felt when they took it. Until 1910 heroin was marketed as a non-addictive substitute for morphine, and as a cough medicine for children. However, heroin was actually much stronger than morphine and even more addictive, with withdrawal symptoms occurring after as little as three days of use. The sale of opiates was first controlled in the United States in 1914, and ten years later the federal government banned the importation, sale and manufacture of heroin on American soil. Heroin has been a proscribed substance in all jurisdictions ever since.

UNTIL 1910 HEROIN WAS MARKETED AS A NON-ADDICTIVE SUBSTITUTE FOR MORPHINE, AND AS A COUGH MEDICINE FOR CHILDREN.

With the worldwide ban on the sale, manufacture and use of the drug – unless it is in strictly controlled ways by registered addicts – most of the heroin currently produced is manufactured in illegal laboratories operated by criminal organisations and terrorist groups. Unfortunately, the process of manufacture, although labour-intensive, expensive and dangerous, is well within the capacities of a competent chemist. Morphine is first extracted from the raw opium and then treated with acetic anhydride for six hours at a temperature of 185°F (85°C). The resulting substance is cooled and purified with alcohol, giving it the appearance of a white crystalline powder. This gives it its common name of 'China White'. When caffeine is added, heroin takes on a brown colouration, which is the most common form of the drug on sale in Europe and America. A darker form of heroin, known as 'black tar', is manufactured in Mexico.

The first major producer and exporter of heroin was China. After the fall of the Qing dynasty (1644–1912), China slowly degenerated into civil war. The Chinese criminal underworld gangs, the notorious triads, aided and abetted by unscrupulous warlords, were the first to exploit the heroin trade. The Second Sino–Japanese war (1937–45) and the subsequent Communist takeover first interrupted and then put an end to China's production and trading of opium and heroin. Poppy cultivation moved southward to the 'Golden Triangle', an area covering parts of Myanmar (Burma), Laos and Thailand, while production moved westward, to mafia-operated labs in Sicily from where it was re-exported to Europe and the United States.

© Public Domain

OPIUM POPPY
This harmless-looking flower (*Papaver somniferum*) is the source of three of the world's most addictive drugs: opium, morphine and heroin. (Illustration by Franz Eugen Köhler from *Köhler's Medicinal Plants*, 1887.)

After the Soviet invasion of Afghanistan in 1979, the border region between Afghanistan and Pakistan quickly became the world's leading production centre for opium. Cultivation peaked in 1999, with 350 sq miles (906 sq km) of poppies grown. When the Taliban came to power, they banned poppy cultivation, reducing production to almost zero within one year. However, after coalition forces invaded in 2001, cultivation quickly returned to its pre-Taliban levels and has continued to increase ever since. Afghanistan now produces the majority of the world's opium, with Myanmar second and Mexico third.

Heroin, which was created as a drug to ease the suffering of addiction, has been one of its greatest causes. The drug can be eaten, snorted, smoked or injected, and the effect will vary with the method of delivery. Addicts report a deep sense of euphoria, relaxation and wellbeing. Unfortunately, this comes at an extremely high cost, both social and personal. The drug itself has known negative effects on the human body, although there are examples of high-profile, long-term users who have lived with their addiction for decades. These are not, however, typical heroin users. The majority of addicts are not wealthy enough to have access to good-quality heroin and medical treatment when they fall ill. Many users quickly become socially dysfunctional and have to resort to criminality to fund their drug habit, making them the targets of both criminal gangs and the authorities. The heroin on sale on the streets is seldom pure and is often cut or contaminated with poisonous substances. Another major risk to intravenous injectors is infection through shared or dirty needles, especially from blood poisoning, HIV and Hepatitis C. If this wasn't enough, the rewards of heroin decrease with use, and the addict requires higher and higher doses to reach the same level of euphoria. This, when combined with the variable purity of an illicit narcotic, means that in order to secure the same fix, a heroin user must expose themselves to the ever greater risk of overdose from an unusually pure batch. Finally, the addict is also at constant risk of extremely unpleasant withdrawal symptoms if he or she does not obtain a repeat dose. Withdrawal can set in as quickly as six hours after the last dose.

WORST OF ALL There are very few inventions in this book that, while they may have caused a good deal of harm, have not in some way, however small, also improved the lot of humanity. Heroin is one exception. Heroin has murdered individuals, slowly destroying their bodies and taking away their humanity; it has brought down entire countries in wars and revolutions; it has been the lifeblood of organised crime in the last hundred years; and it is now the main fundraiser for global terrorism.

PLASTIC NOT-SO-FANTASTIC: THE DEPREDATIONS OF POLYTHENE

FAILING

Never got off the drawing board

Didn't work in practice

Killed its inventor

A commercial failure

Unforeseen consequences

Was used for evil ends

A success born of failure

Main Culprits: Major petrochemical companies, food retailers, consumers

Motivation: Convenience

Damage Done: Petroleum depletion; environmental degradation including the creation of a country-sized patch of rubbish in the Pacific

'I often struggle to find words that will communicate the vastness of the Pacific Ocean … Yet as I gazed from the deck at the surface of what ought to have been a pristine ocean, I was confronted, as far as the eye could see, with the sight of plastic.'

Charles Moore, 'Trashed: Across the Pacific Ocean, Plastics, Plastics, Everywhere', *Natural History Magazine,* **November 2003**

Where would we be without plastic? Plastic is now one of the most commonplace materials around us. Everything from our clothing and footwear, work stations, entertainment systems, to our cars and major appliances are manufactured from one form of plastic or another, or more usually, different plastics in combination. Plastics also fulfil a great many unseen functions in the infrastructure of our cities as pipes, wiring insulation and building materials. Take away plastic and our civilisation would grind to a halt. Or would it? Did our great-grandparents have plastic foodwrap, MP3 players and disposable, but undegradable, shopping bags?

Like the petroleum economy, the plastic economy is a relatively recent phenomenon. Many of the plastics we take for granted today were invented between the two World Wars, and only came into their own in the 1950s. Plastics are synonymous with the throwaway consumer society that has beguiled humanity since the 1960s, and therein lies the main problem. They are cheap to manufacture (until we run out of petroleum; see pp. 21–6) and therefore totally disposable. We manufacture millions of tons of the stuff and most of it ends up in the environment, where certain materials, in particular the one we shall concentrate on in this article, can take up to 1,000 years to degrade.

MAKING MISTAKES

One of our most familiar and versatile plastics, polyethylene, or more popularly, polythene, was first made by accident in 1898 by German chemist Hans von Pechmann (1850–1902). The initial discovery of a white, waxy substance at the bottom of a test tube was not thought to be earth-shattering and was quickly forgotten until two British chemists working for the giant petrochemical firm Imperial Chemical Industries (ICI) repeated Pechmann's experiments in the early 1930s. The commercial production of polythene began in 1939.

Polythene comes in a number of different densities with a wide range of uses. The most familiar use of very-low-, low- and medium-density polythene is in food packaging, plastic bags and plastic wrap; higher-density polythene is found in more resistant products such as water and gas pipes, rubbish bins and food tubs.

Although polythene is a recyclable plastic, until recently very little of the hundreds of millions of tons produced every year was ever collected.

Like plastic wrap and plastic packaging, plastic bags are so cheap that it is still more cost-effective to make new ones than to reuse old ones. As a result, polythene ends up in landfill, in the open environment and in the oceans.

Unlike more recent plastics, polythene is not biodegradable, although it does photodegrade, that is, it breaks down in sunlight. Unfortunately, the process can take several hundred years in dry conditions but up to a thousand years in water, which slows down the process.

The problem of plastic bags dumped in the environment is so serious that many countries in the world have taken measures to limit their use, by phasing out free bags or imposing levies on them. A handful of countries, including Bangladesh, Tanzania, Rwanda and Somalia have banned them outright. With bans in several major European countries and China coming into force in the next few years, the era of that throwaway plastic carrier bag is finally coming to an end. Their legacy, however, is going to stay with us for a good deal longer – several centuries, in fact.

Time was when we wanted to dispose of waste we loaded it on a barge and dumped it at sea. The oceans were so big, we reasoned, that they could absorb anything and everything we threw into them. In the case of biodegradable organic materials, such as wood, paper, sewage and food waste, this is true, but the problem comes from our old friends, plastics. Plastics, in particular light polythene, do not degrade in water, and so they float, following the oceanic currents.

In the late 1980s researchers at the National Oceanic and Atmospheric Administration predicted that a great deal of the non-biodegradable debris that we had been dumping in the ocean since World War II would end up in the North Pacific Gyre, a huge vortex created by the rotation of the earth and ocean currents. Confirmation of its existence was not long in coming, when Charles Moore, a yachtsman and oceanographer, stumbled into it when returning home across the North Pacific after taking part in a sailing race (see quote).

THE GLOBE'S RUBBISH DUMP

Estimates of the size of the 'Great Pacific Garbage Patch' vary from 270,000 sq miles (700,000 sq km) to 5,000,000 sq miles (15,000,000 sq km). According to Moore, 80 per cent of the garbage originated on land and the remainder was dumped from ships at sea, either on purpose or accidentally. Debris travelling from Asia takes about 12 months to reach the patch, while debris from the west coast of America takes about five years. A high concentration of this debris is plastic. In a study in 2001, researchers found that the concentration of plastic debris in parts of the patch had reached one million pieces per sq mile (2.5 sq km).

The problem is not restricted to the Pacific; similar patches exist in the other major oceans and seas. The busiest coastal areas of the Mediterranean revealed in excess of 4,000 items of debris per square mile; 77 per cent of this was plastic, of which 93 per cent was accounted for by polythene carrier bags.

Plastic will degrade over time, but all that happens is that it is broken down into smaller and smaller pieces, which remain suspended in the upper layers of seawater, until they are small enough to be ingested by aquatic organisms. These floating plastic particles, known as 'nurdles', or more romantically 'mermaid's tears', are small enough to be confused with plankton and fish eggs. But any fish, bird or marine reptile unfortunate enough to eat them will either starve, as the nurdles obstruct its digestive tract, or die of poisoning as chemicals leech from the plastic. Larger marine animals can eat bigger items, such as whole plastic bags, which can also cause starvation through the obstruction of their digestive tracts.

DUCK ARMADA

There is precious little to laugh about in the catalogue of dire consequences that have resulted from our addiction to plastic, but just once in a while, there is something that raises a smile. In 1992, a consignment of 28,000 Friendly Floatees children's bath toys, in the shape of beavers, frogs, turtles and ducks, was lost at sea on its way from China to the United States. The cardboard-and-plastic containers in which the toys were packaged quickly disintegrated, allowing the diminutive explorers to begin their Odyssey across the world's oceans. Curtis Ebbesmeyer (b. 1943) and other marine scientists have used the spillage to gain a better understanding of ocean currents and where debris is likely to accumulate and land. It took ten months for the first Floatees to reach the Alaskan coastline, some 2,000 miles (3,200 km) distant from their starting point in the middle of the Pacific. By 1996, they had reached the coast of Washington State. Another flotilla of the diminutive toys travelled back toward Asia, and from there north through the Bering Strait, across the Polar ocean, where they were trapped in the sea ice for several winters before being released into the North Atlantic. Their first ports of call were New England and Canada, but another group is thought to be well on its way to British shores.

LOCATION OF THE NORTH PACIFIC GYRE

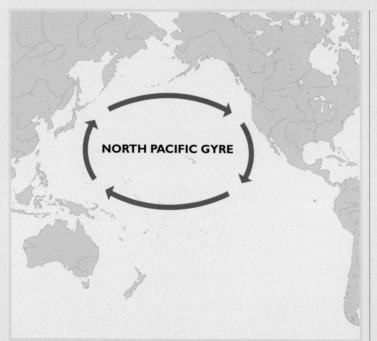

The North Pacific Gyre is a huge vortex created by the rotation of the earth and ocean currents, and it is here that much of the world's discarded plastic ends up. Its size is estimated at between 270,000 sq miles (700,000 sq km) and 5,000,000 sq miles (15,000,000 sq km). It contains an estimated one million pieces of plastic per sq mile (2.5 sq km).

FAILING

Never got off the drawing board

Didn't work in practice

Killed its inventor

A commercial failure

Unforeseen consequences

Was used for evil ends

A success born of failure

GLOWING IN THE DARK: THE DEADLY CRAZE FOR RADIUM

Main Culprit: Marie Curie (1867–1934)

Motivation: Scientific inquiry

Damage Done: Illness and death of Marie Curie and of people manufacturing and using radium products

'Part of what made dial painting an attractive job must have been the work with such a sensational product. The young women applied radium to their buttons, their fingernails, their eyelids; at least one, described by a friend as a "lively Italian girl", coated her teeth with it before a date, for a smile that glowed in the dark.'

Claudia Clark, *Radium Girls,* **1987**

When the double Nobel Prize winner Marie Curie (1867–1934) discovered a new chemical element in 1898, she named it 'radium' from the Latin for 'ray', because it emitted an until then unknown form of radiation. Unfortunately, at the time, no one was aware of the deadly effects of radioactivity on the human body, and the substance is thought to have contributed to Curie's death from aplastic anaemia.

The discovery of the new element spawned a number of crazes in the early twentieth century. The fact that radium glowed in the dark meant that it could be painted onto objects as decoration, and it was used to create luminous watch dials and instrument dials. Radium was added to food, hair creams and toothpastes, and was also used as a constituent of patent medicines, leading to the first deaths from radiation sickness. Another ill-advised application of radioactivity was the 'Fluoroscope', or X-ray shoe-fitting machine, which remained in use in stores in the United States and United Kingdom until the 1960s.

Although radioactivity was first identified in the closing years of the nineteenth century, it was to become the defining discovery of the twentieth, spawning major advances in theoretical physics and huge developments in every other field of science. Three early recipients of the Nobel Prize for physics combined their considerable talents to analyse the phenomenon. The study of the subject began with the discovery of X-rays by the German physicist Wilhelm Röntgen (1845–1923) in 1895. This led a year later to the discovery of radioactivity proper by Frenchman Antoine Becquerel (1852–1908), who was researching the properties of uranium. The third member of this trio of geniuses was Polish-French physicist Marie Curie, who isolated and named several radioactive isotopes, including 'polonium', named for her home country of Poland, and radium.

RADIATING HOPE

Radium (symbol Ra; element number 88 on the periodic table) is a rare element found as a trace element in uranium-bearing ores. Fortunately for us, there are no radium nuggets lying around the landscape, because radium is a million times more radioactive than uranium and is highly carcinogenic. Freshly prepared radium is a brilliant white metal, which quickly darkens when exposed to the air. Radium glows in the dark, giving off a faint bluish-green luminescence. Unaware of the danger she was exposing herself to, Curie handled radioactive elements without taking protective measures, and kept lumps of radium in her desk because she liked its glow. Her work with the element caused her death from a form of leukaemia known as aplastic anaemia. Even today, several of Curie's papers are stored in lead-lined cases and are considered too dangerous to handle without protective clothing.

As with other scientific discoveries, there were those entrepreneurs – deluded or dishonest – who saw an opportunity to make a quick buck. Radium was added to several patent medicines (see pp. 127–30), including Radithor, which was advertised as 'a cure for the living dead' and 'perpetual sunshine'. It took several high-profile deaths until the Food and Drug Administration of America (FDA) moved to ban all radioactive products, which included Radithor, hair creams, food items and toothpastes.

© Corbis

THE CURIES
A man observes an experiment with radium by Marie and Pierre Curie.

However, it was its use as paint for luminescent watch dials that lead to the greatest radium-related death toll. Starting in 1917, the US Radium Corporation manufactured a brand of luminous paints called 'Undark', which was the invention of Dr Sabin von Sochocky. The corporation supplied the US military with luminescent watch dials and instrument faces. Although the managers and scientists at the company protected themselves by using lead screens and protective clothing when handling radium, they took no such care with their employees, who were mainly young women hired to paint the watch dials with the radium paint.

Unaware of the dangers to which they were exposing themselves, the girls used the paint as make-up (see quote) and to decorate their nails, teeth and clothes. At work, they used ordinary paintbrushes to apply a mixture of glue, water and radium to the dials. The brushes quickly lost their shape with use, and the management encouraged the girls to point the brushes with their lips or tongues, allowing the women to ingest huge quantities of the deadly substance. Many of the workers began to suffer from cancers, bone fractures and a necrosis of the jaw later named 'radium jaw'. Hundreds of them died prematurely, but the company denied that the radium was the cause of their illness and organised a medical cover-up. The subsequent court case brought by five 'radium girls' against US Radium Corporation remains one of the cornerstones of labour safety law. And for once, the peddler of death did not escape justice of sorts. In 1928, von Sochocky, the inventor of Undark paint, died of aplastic anaemia.

Anyone who's had a medical or dental X-ray knows how the technician hides behind a heavy lead screen before turning on the X-ray machine, such is the danger from repeated doses of X-rays, and the X-ray itself only lasts a few seconds. Hence the last invention described in this article, the X-ray shoe-fitting machine, will no doubt surprise readers not old enough to remember, but what is sure to appal them is that it remained in use in shoe stores in the United States and United Kingdom until the 1960s. Variously known as the 'Foot-o-scope', the 'Pedoscope', and the 'Fluoroscope', the shoe-fitting X-ray machine was a cabinet-sized wood-and-chrome contraption looking a little like a fairground coin-in-the-slot peepshow machine. The customer placed his or her foot in an opening in the base of the cabinet, and viewing ports in the top of the machine allowed the operator and customer to see an X-ray image of the foot in the shoe.

When a customer put his or her feet into a Fluoroscope, they were effectively standing on top of an unshielded X-ray tube. The machine allowed the operator to select one of three different intensities: high for men, medium for women, and low for children. Exposure time could be set from 5 to 45 seconds, though the average seems to have been 20 seconds. The machines were in use from the 1920s to the 1960s, and while there are no records of the health damage they caused, this must have been significant for the staff of the shoe stores who were in daily contact with the machines and put their hands inside to adjust the shoes.

IF THE SHOE FITS

FAILING

Never got off the drawing board

Didn't work in practice

Killed its inventor

A commercial failure

Unforeseen consequences

Was used for evil ends

A success born of failure

A RECIPE FOR DISASTER: FAST FOOD

Main Culprits: Fast-food chains, consumers

Motivation: Greed and gluttony

Damage Done: A worldwide obesity epidemic; health problems including increasing rates of heart disease, type-2 diabetes and mad cow disease

'Americans now spend more money on fast food than on higher education, personal computers, computer software, or new cars.'

Eric Schlosser, *Fast Food Nation*, 2001

Since the publication of Eric Schlosser's *Fast Food Nation* (2001) and the screening of Morgan Spurlock's *Super Size Me* (2004), it's been fashionable to take a pop (no pun intended) at fast food as the main cause of the current obesity epidemic that has overtaken the developed world and is fast spreading to the developing world. However, the concept of fast food is not itself to blame.

Fast food has existed since antiquity as a conveniently packaged, portable meal that can be eaten at work, in the fields, on the road, and at fairs and festivals. All food cultures have their own versions of fast food: Neapolitan pizza, which started life as a slice of bread flavoured with garlic, olive oil and tomato purée – hold everything else; the British sandwich – first made, so legend has it, for the fourth Earl of Sandwich (1718–92) one evening when he was playing cards; the Cornish pasty, a pastry case filled with meat and vegetables eaten by tin miners in the west of England; the hot dog, traced back to Germany in 1480 when it was given to the citizens of Frankfurt as a coronation treat; and the hamburger, claimed by half a dozen American chefs at the end of the nineteenth century.

If fast food has contributed in a significant way to the obesity crisis – and there are few people today who could convincingly argue that it has not – it isn't the idea of convenience food that's the cause but the interplay of social, economic and dietary factors that have combined to transform something healthy, tasty and convenient into something physically and socially toxic. This article will concentrate on the hamburger-fries-and-soda combo because it is the most popular form of commercialised fast food available today, tracking its origins from somewhere small and folksy in the United States in the late nineteenth century, to its current incarnation as the mainstay of the multi-billion-dollar global fast-food behemoth.

FAT FOOD

A braised ground-beef patty in a bun – what could be simpler and more convenient? It's a wonder that no one had thought of it before 1885, the date claimed by the first two rival contenders for the title of inventor of the hamburger: Charles Nagreen of Seymour, Wisconsin, who flattened a beef meatball between two slices of bread to make it easier to eat at that year's Outagamie County Fair; or the Menches Brothers, who also made the first 'hamburger' at the Eerie County Fair, Hamburg,

New York, when they ran out of pork fillings for their sandwiches. They christened their creation the 'Hamburger sandwich'. Yet a third contender was Fletcher Davis from Athens, Texas, who made no claims of his own but is said to have made ground-beef sandwiches in Athens in the late 1880s and sold them at the St. Louis World's Fair of 1904. The likely truth is that they, and a couple of other claimants from the 1900s, probably all got the idea separately, and the recipe quickly spread from county fair to world's fair, from diner to concession stand; it was simple, homely, healthy, inexpensive, easy to make and easy to eat on the move.

Now that we have discovered the origins of fast food's bestselling product, we can move on to the next chapter of the story: the development of the fast-food outlet. A fast-food outlet is quite unlike a conventional sit-down, table-and-chair, waiter-service restaurant. It's a cross between a concession stand, a bar and a diner. The customer comes in, orders his or her food at the counter, and eats it there on functional seating or takes it out. The first diner, which opened its doors in Providence, Rhode Island, operated from a horse-drawn wagon from 1872, taking meals to the factory gate. The next significant development was the Automat chain of restaurants, which opened its first branches in Philadelphia, Pennsylvania, in 1902, and in New York City in 1912. The Automat was the first true fast-food outlet, though instead of a counter with burger hops, the dishes were dispensed from coin-operated vending machines separating the customers from the kitchen area. Finally, in Wichita, Kansas, in 1921, Billy Ingram and Walter Anderson founded the White Castle Co., which was the first chain of low-cost, high-volume hamburger restaurants in the world. White Castle's major contribution to fast food was to allow the customers to see their patties being griddled.

THE FIRST DINER, WHICH OPENED ITS DOORS IN PROVIDENCE, RHODE ISLAND, OPERATED FROM A HORSE-DRAWN WAGON FROM 1872, TAKING MEALS TO THE FACTORY GATE.

All the ingredients of the fast-food industry had finally been brought together: the hamburger, the ordering and serving system, and strong corporate branding. But before fast food could really take off in the United States, several important social changes had to take place. The first was the development of the automobile culture, which took off after World War I, when Ford made cars truly affordable; the second was suburbanisation; and the third was the emancipation of women.

Suburbanisation moved people away from the city centres where the traditional markets, food shops and eateries were concentrated, enabling a new type of restaurant, catering to a more mobile customer base, to develop. And the emancipation of women meant that working mothers had less and less time to shop and cook for their families, who were more likely to eat out.

Say fast food, and one name – or should that be one outsized letter – naturally comes to mind. McDonald's, however, had very humble origins. The original McDonald's, named for its owners, brothers Richard 'Dick' (1909–98) and Maurice 'Mac' (1902–71), was a drive-in barbecue shack, which they opened in San Bernardino, California, in 1940. In 1948, the brothers reinvented the restaurant, offering a simplified menu of burgers, fries, shakes, coffee and Coca-Cola, served in disposable packaging. They introduced the 'Speedee Service System', which applied Henry Ford's production-line methods to food preparation. In 1953, they franchised two further McDonald's restaurants, the first in Phoenix, Arizona, and the second in Downey, California. Their success attracted the attention of their milkshake-machine supplier, Raymond Kroc (1902–84). Kroc immediately grasped the enormous potential of the McDonald's concept, and he went into business with the brothers in 1954. The relationship was not an entirely happy one, because Dick and Mac did not want a nationwide chain of franchised restaurants bearing their name. In 1961, Kroc bought the brothers out for US $2.7 million – surely the deal of the millennium – including the rights to the McDonald's name.

THE BIRTH OF THE BIG MAC

McDonald's is now the flagship of the fast-food industry. In 1998, it had around 3,000 restaurants outside the continental United States; today it is the world's largest food retailer with 31,000 outlets in 119 countries, employing 1.5 million staff, and feeding an estimated 47 million happy eaters every single day. The busiest McD's is in Moscow, Russia, and the largest McDonald's restaurant is in Beijing – an interesting footnote to the Cold War, to be sure. But McDonald's is only the tip of a fast-food iceberg of truly humongous proportions toward which humanity, aboard the good ship RMS *Titanic*, is sailing full steam ahead (please pass the ketchup). Of McDonald's competitors, Burger King has more than 11,100 restaurants; Wendy's has 6,500; KFC is to be found in 25 countries; Subway has 29,186 restaurants in 86 countries; Pizza Hut is in 26 countries; and Taco Bell has 278 restaurants in 12 countries outside the United States. In 2006, the global fast-food market was worth US $102 billion, representing a volume of 80 billion transactions. The major growth is in the developing world, so long denied the benefits of the Big Mac. In predominantly vegetarian India, for example, the fast-food industry is growing at an annual rate of 40 per cent.

Two of fast food's most trenchant and best-known critics are journalist Eric Schlosser, who wrote *Fast Food Nation* (2001), and documentary filmmaker Morgan Spurlock, who directed and starred in *Super Size Me* (2004). Schlosser's book is an exposé of the manufacturing and marketing practices of the industry. He highlights labour and health issues, including the spread of bovine spongiform encephalopathy (BSE), better known as 'mad cow disease', and the targeting of children by the fast-food industry through advertising and subsidised school catering. Spurlock went for a much more hands-on approach: he famously went on the 30-day 'McDiet', eating exclusively at the golden arches three times a day, and taking only the minimum daily recommended amount of exercise. His average calorie intake of 5,000 kcal (approximately twice what he actually needed) increased his bodyweight by 24.5 lb (11.1 kg) and his body mass index by 13 per cent. He and his medical team reported a wide range of other psychological and physiological effects from the 'McDiet', ranging from liver problems to sexual dysfunction.

WHOLLY BAD | The fast-food industry has become a bit like a drug dealer at the school gates, pushing unhealthy products full of trans fats, high-fructose corn syrup and additives (see pp. 160–64) to adults and children alike. They have made it extremely cheap and extremely tasty – with all the fats, sugars and flavourings. We have bought into the idea like pigs at the trough and, surprise, surprise, the net result has been the obesity epidemic with its many associated health problems.

Several critics have compared the behaviour of the fast-food majors now with that of the tobacco companies in the 1970s and '80s. But, hello, brothers and sisters, food is not a physiologically addictive substance like nicotine (see pp. 32–7) or heroin (see pp. 136–40). No one is driven screaming through the night to McDonald's, Taco Bell or Pizza Hut by an irresistible craving for a fix of monounsaturated fat or high-fructose corn syrup. If the fast-food industry must shoulder the responsibility of making and selling cheap and tasty edible junk, then surely we have to accept our share of the blame for buying and eating the stuff.

Fast food is a classic example of a particular kind of unforeseen consequence: the combination effect. Unlike some of the discrete 'bad' inventions that we have looked at, such as arsenic (see pp. 44–8) or switchblades (see pp. 75–8), fast food consists of a large number of different innovations – the hamburger, the diner concept, potato fries and so on – that individually do not have particularly marked detrimental effects, but when combined have very serious negative consequences for human wellbeing. It is really a case of the whole being a whole lot worse than the parts. And therein lies the major problem about finding a solution to the problems created by fast food. While legislators can ban the sale and manufacture of noxious chemicals and criminalise the carrying of weapons in public, they are never going to outlaw the all-too-popular hamburger, fries and soda combo.

PREVALENCE OF OBESITY AROUND THE WORLD

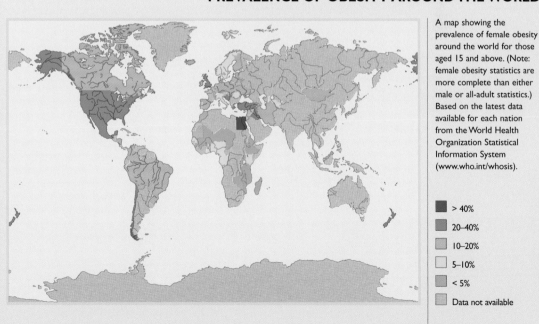

A map showing the prevalence of female obesity around the world for those aged 15 and above. (Note: female obesity statistics are more complete than either male or all-adult statistics.) Based on the latest data available for each nation from the World Health Organization Statistical Information System (www.who.int/whosis).

- > 40%
- 20–40%
- 10–20%
- 5–10%
- < 5%
- Data not available

FAILING

Never got off the drawing board

Didn't work in practice

Killed its inventor

A commercial failure

Unforeseen consequences

Was used for evil ends

A success born of failure

THE FIRST WMD: TESLA'S TELEFORCE

Main Culprits: Nikola Tesla (1856–1943) and his disciples

Motivation: Scientific exploration

Damage Done: Encouragement of pseudo-science and crank theories

'My invention requires a large plant, but once it is established it will be possible to destroy anything, men or machines, approaching within a radius of 200 miles [321 km]. It will, so to speak, provide a wall of power offering an insuperable obstacle against any effective aggression.'

Nikola Tesla, *A Machine to End War*, 1937

Since the 1950s, the peace of the world, such as it is, has been assured by mutually assured destruction (MAD) – the idea that the only possible outcome of a nuclear conflict between NATO on one side and Russia or China on the other would be the annihilation of the entire planet. In this no-win situation, we can only hope that no one will ever push the red button.

In the 1930s, however, the brilliant Serbian electrical engineer Nikola Tesla (1856–1943) proposed something a little more sophisticated and slightly saner, had it worked. His invention was not a bomb but an incredibly powerful high-energy beam weapon, which he called 'Teleforce' but was quickly renamed the 'death ray' by the press of the day. Tesla argued that Teleforce would make conventional weapons obsolete overnight, and thus make war pointless. Tesla's idea was the ultimate Strategic Defense Initiative (SDI), half a century before President Ronald Reagan (1911–2004) came up with a very similar idea for use against the Soviets in the 1980s.

The best description of Tesla's Teleforce can be found in a 1937 *New York Times* interview, in which Tesla described how a country could be protected by a ring of towers, each one housing a Teleforce generator. An enemy approaching by land, sea or air would be instantly detected, and each tower, in the words of the *Times*, 'will send concentrated beams of particles through the free air, of such tremendous energy that they will bring down a fleet of 10,000 enemy aeroplanes at a distance of 250 miles [400 km] from the defending nation's border and will cause armies of millions to drop dead in their tracks.' Tesla offered his death ray to the British for US $3,000,000, promising to make the British Isles invulnerable from attack. Whitehall opted for radar and spitfires instead. He then tried to sell the death ray to the League of Nations, without success, but he actually managed to obtain US $25,000 from Stalin (1878–1953), without ever delivering anything of use to the Soviets.

REMOTE-CONTROLLED PEACE

Teleforce was not a laser beam or a weapon firing a beam of subatomic particles such as protons or electrons, as neither of these technologies had been perfected at the time. Tesla's device was based on a giant Van de Graaff electrostatic generator combined with an open-ended vacuum tube, which together would accelerate a beam made up of minute particles of tungsten or mercury to a velocity of about 48 times the speed of sound. In other words, Teleforce was a very sophisticated cannon. Tesla gave the death ray a range of 250 miles (400 km), although he claimed that the ray could operate over much greater distances without loss of power or accuracy. In 1940 he estimated that each station would cost no more than US $2,000,000 to build and that a chain surrounding the United States could be constructed in a little more than three months. However, building the weapon depended on solving what turned out to be a series of insuperable problems, including generating the huge electrical charge necessary and creating a working open-ended vacuum tube through which the beam could be fired.

Although with hindsight, Tesla's claims for his Teleforce technology strike contemporary scientists as half-baked, at the time they were taken seriously because of his many scientific achievements in the fields of electrical engineering and radio. His patents and theoretical work formed the basis of modern alternating current (AC) electrical power systems. His contributions were so important that contemporary biographers of Tesla have called him 'the man who invented the twentieth century' and 'the patron saint of modern electricity'. After he demonstrated wireless communication (radio) in 1891, he was hailed as America's greatest electrical engineer. Congress even declared that it was Tesla and not Marconi who was the inventor of radio technology. However, Tesla's eccentric personality, coupled with his sometimes unbelievable, bizarre claims about possible future scientific and technological developments, led to his ostracism from the scientific community. He died impoverished and disgraced at the age of 86. In 1960, his many contributions to electrical science were posthumously recognised when the SI unit of magnetic flux density, or magnetic induction, was named the tesla in his honour.

STAR WARS

Although Teleforce proved to be a scientific fantasy, it was the first in the long line of proposals for high-energy particle weapons that culminated in President Reagan's Strategic Defense Initiative in 1983. The SDI envisaged a layered ground- and space-based defence system that would knock out Soviet nuclear missiles before they had a chance of reaching American soil. The space component gave the initiative its more popular name of 'Star Wars'. The weapons, to be controlled by the mother of all supercomputers, included space- and ground-based nuclear X-ray lasers, subatomic particle weapons and electromagnetic rail guns, which would be guided by a network of space-based sensors and mirrors used to target the weapons toward the incoming Intercontinental Ballistic Missiles.

Unfortunately, there were just a few technical, military and political flaws in the plan. On the technical side, several of the proposed weapons were, and still are, on the drawing board, and several others were no more realisable than Tesla's Teleforce. Another major problem was that the system was not designed to protect against nuclear weapons that were not launched on ICBMs into high orbit. A cruise missile, for example, evades detection by hugging the surface of the planet, and it, too, can carry a large nuclear payload. The use of this type of non-conventional nuclear missile would have made the SDI useless.

© Public Domain

Even if the technical issues had all been overcome, there remained a good many military and political issues facing the SDI. The United States' major adversary during the Cold War, the USSR, ceased to exist in 1991, and a similar threat from a global superpower has failed to emerge elsewhere in the world.

Star Wars, which was designed to detect and repulse an attack with space-faring missiles from a rival superpower, is useless against the United States' current foes, such as Al-Qaeda and the Taliban, who favour much more low-tech terror tactics.

NIKOLA TESLA
The erratic genius who is considered to be the 'patron saint of electricity'.

FAILING

Never got off the drawing board

Didn't work in practice

Killed its inventor

A commercial failure

Unforeseen consequences

Was used for evil ends

A success born of failure

A LITTLE BIT OF WHAT YOU FANCY: TRANS FAT, HFCS AND FOOD ADDITIVES

Main Culprits: Food and drinks manufacturers

Motivation: Greed

Damage Done: Obesity and a host of different health issues

'Scientific evidence shows that consumption of saturated fat, trans fat, and dietary cholesterol raises low-density lipoprotein (LDL), or "bad cholesterol", levels, which increases the risk of coronary heart disease (CHD). According to the National Heart, Lung, and Blood Institute of the National Institutes of Health, more than 12.5 million Americans have CHD, and more than 500,000 die each year.'

US Food and Drug Administration, 'Revealing Trans Fats', 2003

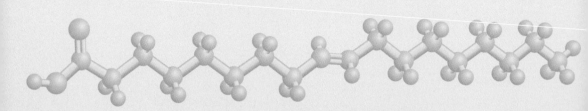

In the article on fast food (pp. 150–55), we looked at the origins and development of the fast-food industry, and of its bestselling product, the hamburger. We mentioned the most obvious health effect of a diet heavy on fast food: obesity. However, space did not allow us to examine in detail the ingredients that make fast food so tasty, convenient and cheap, and at the same time can cause a range of health problems. The fast-food business model of a high turnover of standardised products at the most competitive price means that manufacturers have industrialised the production of their foodstuffs – burgers, fries, shakes and so on – and at the same time found the cheapest ingredients. There is only so much cost-cutting you can get away with, with meat and buns, but a great deal of scope offered by fats, sweeteners and additives to make cheaper ingredients more attractive and palatable. This article will look at three main areas of concern: trans fat, high-fructose corn syrup (HFCS) and colour- and flavour-enhancing additives, which have all been implicated in serious health problems.

LIVING OFF THE FAT

Trans fat has a scientific yet faintly sinister ring to it. If you were to talk about good old-fashioned honest fat, you'd know where you were. It's not that good for you, or at least too much of it isn't, but at least it's natural. For most of human history, fat was extracted from certain plants, such as the olive, or made from animal products, such as buttermilk. Northern Europe, in contrast to olive-rich southern Europe, always depended on animal fats such as butter and lard, a practice that it exported to North America during the colonial period. However, animal fat is expensive, and it also quickly goes rancid unless refrigerated. Just over a century ago, the work of two chemists, the Frenchman Paul Sabatier (1854–1941) and the German Wilhelm Normann (1870–1939), led to the invention of fat hydrogenation; that is, the process that turns liquid oil into solid fat. In 1909, Procter & Gamble bought the US rights to the process, and in 1911 they began marketing the first hydrogenated 'shortening', Crisco.

Hydrogenation also made margarine, a product invented in France in 1873, into a much more marketable product in the United States. The main draw of hydrogenated fats was the price, as the new solid fats were made from much cheaper vegetable oils; they also had a longer shelf life; and they were found to have better baking properties than the common traditional animal shortening, lard. The low price point made trans fat particularly attractive to the food industry, and in the past few decades they have replaced more expensive fats in processed meals, fast food, snacks and baked goods.

Trans fats are not entirely unknown in nature, and humans have always been exposed to them in their diet. Animal fat contains between two and five per cent, although this natural form of trans fat is chemically different from the artificial product, and breaks down differently in the body. In any case, until the 1960s, the high cost of animal products ensured that most people ate only moderate amounts of trans fat in their diet. Modern processed food, however, can contain up to 40 to 45 per cent of trans fat; vegetable shortenings contain around 30 per cent trans fat; and margarines up to 15 per cent. Historically, fast-food chains have used trans fats in all their products, and have only phased them out after the introduction of legislation limiting or banning their use in many developing countries. According to the US National Academy of Science, trans fats of any origin are 'not essential and provide no known benefit to human health'. Quite to the contrary, by raising 'bad' cholesterol (see quote), they are estimated to cause an annual death toll of between 30,000 and 100,000 from coronary heart disease (CHD) in the United States alone. In addition to CHD and obesity, trans fats have been implicated in increased risks of a number of diseases, including Alzheimer's, breast cancer, diabetes, liver disease and infertility. Burger and fries, anyone?

Along with salt, the most common flavouring agents used in food production are sweeteners. The original sweetener is honey, which is about as natural as an ingredient can get but, historically, humans have refined their sugar (sucrose) from sugar cane and beet. In 1927 a group of chemists discovered how to use enzymes to increase the fructose (another type of sugar found in fruit) content of corn syrup (a type of starch), turning it into high-fructose corn syrup. The manufacturing process was refined and commercialised by Japanese scientists in the mid-1960s, and the new product found a ready market in the US food industry in the 1970s. HFCS comes in several formulations: HFCS 90 is 90 per cent fructose and 10 per cent glucose; HFCS 55 is 55 per cent fructose, 45 per cent glucose, and HFCS 42 is 42 per cent fructose

HISTORICALLY, FAST-FOOD CHAINS HAVE USED TRANS FATS IN ALL THEIR PRODUCTS, AND HAVE ONLY PHASED THEM OUT AFTER THE INTRODUCTION OF LEGISLATION LIMITING OR BANNING THEIR USE IN MANY DEVELOPING COUNTRIES.

and 58 per cent glucose. HFCS has several advantages over sugar in that current import tariffs for cane sugar make it cheaper in the United States, and that because it is liquid it is easier to transport and blend. As a result most soda manufacturers (see pp. 70–74) in the United States use HFCS rather than cane sugar.

Calorifically, both sucrose and solid HFCS weigh in at the same 4 kcal/gram. Natural sucrose is actually made of glucose and fructose, so is there really a problem here, or is this just some more health-food freakery at work? Well, according to a growing body of evidence, there is. But first it must be said that we in the developed world are massively over-consuming sugar – sucrose, glucose, fructose or HFCS – in our diet. We eat far too much of it and we're getting fatter as a result. However, HFCS may just add that little extra to the equation. Sucrose is a molecule consisting of glucose and fructose bonded together, while in HFCS the sucrose and glucose are separate. When we digest sucrose, the enzyme sucrase breaks down the sucrose into glucose and fructose so the body can absorb them at a regulated rate. With HFCS, the body can absorb them directly all in one go – the result: sugar spikes, obesity, insulin resistance and diabetes. The industry response has been rather predictable. One HFCS manufacturer funded research that concluded: 'HFCS does not appear to contribute to overweight and obesity any differently than do other energy sources.'

ADDITIVE ADDICTION

Time was when most of the flavour came from the food itself, and all that you added to it was a little salt, a little sugar or a handful of spices. Colour-wise, it either kept its natural tint or went a shade of brown. Not so today in the world of processed and fast food, where the appeal is first to the eye, then the nose and only later the palate. There are now literally hundreds of permitted food additives, including acids, acidity regulators, anti-caking agents, anti-foaming agents, antioxidants, bulking agents, colourings, flavours, flavour enhancers, humectants, preservatives, stabilisers, sweeteners and thickeners. Several food additives, both natural and artificial, are now subject to bans because they have been discovered to be toxic or carcinogenic (see Arsenic, pp. 44–8), but recent research indicates that there may be a link between certain artificial additives and Attention Deficit Hyperactivity Disorder (ADHD) in children.

It may very well be the case that singly, a small amount of trans fat, a few spoonfuls of HFCS and the odd pinch of food additive won't do the average person any harm. But in the modern world we are not talking about the occasional consumption of small quantities of these substances, but the regular consumption in large amounts of all three. Trans fat is being phased out, and HFCS and many food additives will probably follow soon. Unfortunately, we have a couple of generations who grew up on them who will have to live and die prematurely with their own and the food industry's choices.

Undoubtedly the worst aspect of the cynical debasement of the quality of our foodstuffs for profit is its current and future impact on the health and wellbeing of children. Adults, even if they choose to abdicate all sense of personal responsibility for their diet, do have freedom of choice to eat well or badly. There can't be a single adult alive in the developed world who doesn't know that too much fat and sugar is bad for you. Children, however, eat what they are given by their parents, or are brainwashed to demand the products that they see advertised on TV – usually not the 'fresh-fruit-and-water' option but HFCS, fat and additive-packed sodas, snacks, biscuits and sweets. An adult choosing to eat him- or herself to an early grave is tragic, but encouraging a child to do the same must be criminal.

COLOUR ME WELL: SPECTRO-CHROME THERAPY

FAILING

Never got off the drawing board

Didn't work in practice

Killed its inventor

A commercial failure

Unforeseen consequences

Was used for evil ends

A success born of failure

Main Culprit: Dinshah Ghadiali (1873–1966)

Motivation: Greed

Damage Done: Misled and swindled the gullible; caused deaths of patients who did not seek proper medical treatment

'When it will be known that you are using Spectro-Chrome, you may probably receive adverse advice from opponents, Medical Doctors or otherwise. Take our advice and listen to them not. Spectro-Chrome – In Every Home, means the crumbling of their age-old moth-eaten doctrines and the upholding of the Torch of Emancipation, releasing you from their orthodox and autocratic grasp. For instance, they will tell you to stop the eating of all Starches and Sugars and inject Insulin, because you have "Diabetes". You ask for our FREE GUIDANCE and we shall tell you: "Stop Insulin at once and irradiate yourself with Yellow Systemic alternated with Magenta on Areas 4 or 18 and eat plenty of Raw or Brown Sugar and all the Starches!!!"'

Dinshah Ghadiali, *Family Health Protector*, 1943

The idea that colours influence the mind, spirit and body is as old as human civilisation. In traditional Indian thought, the seven *chakra* (Sanskrit for 'wheel'), or energy centres of the body, are each ascribed a different colour – though there is no agreement between various Indian systems. Colour is said to influence mood, and marketers now spend considerable amounts of money picking the right colours for packaging and the interior-design schemes of stores and restaurants. However, the meanings of colour are socially defined, and different cultures have very varied readings for the same shade. In Asia, for example, yellow and not purple is the colour of royalty, and white and not black is the colour of mourning. Colour had applications in the pre-modern medical systems of ancient Egypt, India, China and ancient and medieval Europe, but none of these claims have withstood the era of scientific testing. A complete lack of scientific evidence, however, has never stopped medical quacks and pseudo-scientists; on the contrary, it liberates them to make ever more outlandish claims. Charlatans do not appeal to our rationality but play on our hopes and fears – our fears of illness and death and our hopes that a painless magic bullet, patent medicine or therapy will instantly make everything better.

A COLOUR CURE

In 1911, a 28-year-old man of what appeared to be a truly extraordinary intellectual pedigree, Dinshah Ghadiali (1873–1966), emigrated from Bombay (now Mumbai), India, to New Jersey, United States. In his autobiography, Ghadiali claimed to have graduated high school at the age of eight, and taught college mathematics at the age of 11. His academic credentials included a PhD, a doctorate of 'legal' law, as well as several medical degrees – without, however, providing any evidence of having attended any university or medical school. To these imaginary qualifications, he later added several grand-sounding titles, including 'Fellow and Ex-Vice-President, Allied Medical Associations of America', 'Member and Ex-Vice-President, National Association of Drugless Practitioners', and 'President, American Association of Spectro-Chrome Therapists'.

Ghadiali's theory of 'spectro-chrome-metry', which he unveiled to the world in 1920, was a curious combination of traditional Indian thought and modern pseudo-scientific theory and terminology. He claimed that the elements that made up the human body corresponded to different colours: oxygen with red, hydrogen with blue, nitrogen with green, and carbon with yellow – an echo of the four 'humours' common to many ancient medical systems. In a healthy individual, the colours were balanced but disease caused an imbalance in the 'Radio Emanations of the Chemical Body, called the Aura or the Auric Vehicle'.

In order to restore balance and health, Ghadiali devised the spectro-chrome, a light box containing a 1,000-watt light bulb with an opening that could be covered by 'Attuned Colour Wave Slides'; that is, filters made of coloured glass. The filters could be used singly or in combination to produce a total of twelve colours, each claimed to have a particular curative or stimulating property.

In the book accompanying the machine, Ghadiali described the functions of the different colours. Red, for example, energised the liver and stimulated red blood cell production, while lemon was a bone builder. He prescribed spectro-chrome therapy for all ailments except broken bones, sometimes with disastrous consequences (see quote). At one of his many trials, in 1945, a prosecution witness described how Ghadiali had instructed him to shine yellow light on his father, who had just lapsed into a diabetic coma. The man, whose life could have been saved by the administration of a small amount of carbohydrate, then died bathed in the warm, useless glow of the spectro-chrome.

Ghadiali was no stranger to the law. In 1925 he was arrested and tried for transporting a 19-year-old girl from Oregon to New Jersey for immoral purposes. He served four years in prison, but only two years after his release was convicted again and fined in Ohio for the fraudulent claims that he had made about his spectro-chrome devices. In 1945, the US Food and Drug Administration moved against him, and he was found guilty, fined US $24,000 and sentenced to five years' probation. Undeterred, Ghadiali set up the Visible Spectrum Institute to continue peddling his colour therapy. Finally, in 1959, the government obtained a permanent injunction to prevent Ghadiali from promoting spectro-chrome in any shape or form.

He died a rich man, having made an estimated US $1 million from the gullible and the desperate. To the end, he claimed to be a misunderstood genius persecuted by a jealous and vindictive 'Medical Octopus'. After

GHADIALI WAS NO STRANGER TO THE LAW. IN 1925 HE WAS ARRESTED AND TRIED FOR TRANSPORTING A 19-YEAR-OLD GIRL FROM OREGON TO NEW JERSEY FOR IMMORAL PURPOSES.

Ghadiali's death in 1966, his family continues to manage the business as a non-profit, scientific, educational, tax-exempt organisation, selling Ghadiali's books and courses on spectro-chrome therapy.

NEW AGE, OLD COLOURS

Colour therapy, also known as 'colourology' or 'chromotherapy', has known something of a revival as one of the complementary New Age therapies. It links up with other popular New Age beliefs about the healing powers of crystals and the existence of bodily auras, the colour of which change with mood and health. A trawl through the Internet quickly brought up this explanation of today's colour therapy:

> The energy relating to each of the seven spectrum colours resonates with the energy of each of the seven main chakras of the body. Balance of the energy in each of the body's chakras is very important for health and wellbeing. Colour therapy can help to rebalance and/ or stimulate these energies by applying the appropriate colour to the body and therefore rebalance our chakras.

Happily for us, the law forbids these latter-day spectro-chromists from making any medical claims for their dubious practices. Proper medical trials, using control groups of patients, have proved over and over again that the major effect demonstrated by most complementary therapies is not the mystic power of crystals, auras, foot massages, hand waving, chanting or essential oils carefully massaged into acupuncture points or chakras, but the placebo effect – the power of the mind and body to heal itself. Unhappily, there will always be enough people desperate for a little bit of hope parcelled up in a lot of mumbo-jumbo to allow the Ghadialis of this world to make a living.

THAT HEAVY FEELING: LEADED PETROL AND PAINT

FAILING

Never got off the drawing board

Didn't work in practice

Killed its inventor

A commercial failure

Unforeseen consequences

Was used for evil ends

A success born of failure

Main Culprit: Thomas Midgley Jr (1889–1944)

Motivation: Scientific inquiry

Damage Done: Lead poisoning on a planetary scale

'Ask a historian to name the people who have changed our world in the twentieth century. He or she will probably name some world leaders, major thinkers whose ideas have had an impact, some scientists, some inventors and perhaps some creators of popular culture. It is unlikely that the name of Thomas Midgley Jr will feature on any list. Yet in a very real sense Thomas Midgley changed the world in a way that no statesman or major thinker did. These merely changed the lives of a lot of people who live in the world. They did not leave their mark on the planet in the way Thomas Midgley left his.'

Norman Moss, *Managing the Planet*, 2000

A form of the metallic element lead (Pb), known as tetra-ethyl lead (TEL) was first used as an additive for petrol in the early 1920s to prevent engine 'knocking'. It remained in common use in the United States until the 1980s and in Europe until the 1990s. At the same time, domestic paints using lead as a pigment or to increase its durability were used until the late 1970s in American and European homes. Combined with other sources of lead released into the environment by human-made products, leaded petrol and paints have been responsible for a range of disorders including neurological, cardiovascular, reproductive and renal disease. Another unforeseen consequence of high levels of environmental lead pollutants is the crime wave that afflicted many of the United States' major urban centres in the 1980s.

THAT LEADEN TOUCH

There are some inventors – the da Vincis, Brunels and Edisons of this world – who, in spite of the occasional failure as chronicled in these pages, seem to have the Midas touch. They are the inventors' inventors. They have changed the world, mostly for the betterment of humanity. But equally there are those who could be described as 'anti-inventors', because their inventions, while at first holding great promise, turn out to be complete disasters for humanity and for the planet as a whole. One such was the American chemist Thomas Midgley Jr (1889–1944), a man described as having 'had more impact on the atmosphere than any other single organism in earth history'. Midgley's first claim to infamy was the discovery of a substance to prevent knocking in internal combustion engines (see pp. 108–12). Engine knocking occurs when the mixture of petrol and oxygen in the piston chamber does not combust smoothly but 'detonates'. At best knocking increases engine wear; at worst, it can punch holes into the piston casing or piston head. What was required was a substance that would even out combustion. The substance Midgley stumbled across in 1921 while he was working for the US chemical giant DuPont was tetra-ethyl lead $((CH_3CH_2)_4Pb)$.

Tetra-ethyl lead is a colourless, odourless liquid that dissolves in petrol. When added to the mix, it prevents the ignition of unburned petrol and oxygen from igniting during the exhaust stroke of the engine cycle. Those readers old enough to remember leaded petrol will recall the different 'octane' ratings of fuels, which was a measure of how much tetra-ethyl lead was present. The higher the octane, the better the performance of the engine; unfortunately, it also meant the greater the amounts of noxious waste products released into the environment.

Tetra-ethyl lead production began at a DuPont plant in Dayton, Ohio, but the process was so dangerous that cases of lead poisoning and deaths occurred immediately. In 1924 the General Motors Chemical Company (GMCC), with Midgley installed as vice-president, took over production at a plant in New Jersey with an even more dangerous manufacturing process. Predictably, the death toll continued to rise, and the state authorities moved to close the plant and forbade production of tetra-ethyl lead in New Jersey. Midgley himself had to take extended leave because of lead poisoning, which he himself had made worse by publicly inhaling tetra-ethyl lead to demonstrate how safe it was.

There is even a story that he died from lead poisoning, but this is wishful thinking on the part of some overzealous environmental campaigner. In 1940 Midgley contracted polio. Paralysed, he designed a device using pulleys and ropes to lift himself in and out of bed. He died in 1944 after getting entangled in the ropes and suffocating. We shall return to the unfortunate Mr Midgley, the man whose touch turned everything to lead, in a later entry (see CFCs, pp. 193–7).

© Corbis

THOMAS MIDGLEY
The winner of the dubious accolade of being the single organism to have had the most detrimental effect on the atmosphere in history.

Lead, along with gold and silver, was one of the first metals to be mined and used by humans. It is common in nature, easy to extract, and is soft, pliable and easily worked. The most common ancient uses included construction, pottery glazing, lead slingshot and weights. Historically, various compounds of lead have been used as pigments, in particular white (lead carbonate) and yellow (lead chromate). In addition to the anti-knocking agent tetra-ethyl lead and pigments, the modern uses of lead include plumbing products, solder, batteries, children's toys, ceramics and electrical insulation.

The toxicity of lead has been known since ancient times, and the Romans, who were among the greatest users of the metal – as an ingredient for cosmetics, medicines, as a food preservative and to make their drinking vessels and water pipes – were also among the greatest sufferers of lead poisoning. One historian has theorised that so many of the Roman emperors became senile or insane homicidal maniacs because of the amount of lead they absorbed. Lead attacks the peripheral and central nervous system of humans and animals. Children under the age of 12 are particularly susceptible, and research has confirmed a direct link between severe learning disabilities and exposure to environmental lead.

The symptoms of lead poisoning include neuropathy (loss of feeling and paralysis in the extremities), abdominal pain, insomnia, lethargy or hyperactivity and, in acute cases, seizures and death. Other associated effects are anaemia, kidney disease and reproductive problems. Adults are exposed to lead through their work, but for children, the main sources of poisoning are from eating contaminated soil and from breathing in and eating lead dust or chips from flaking lead-based paints.

ONE OF THE MOST INSIDIOUS AND DANGEROUS CAUSES OF LEAD POLLUTION WAS MIDGLEY'S TETRA-ETHYL LEAD. WHEN HIGH-OCTANE PETROL IS BURNED, MINUTE PARTICLES OF LEAD ARE RELEASED INTO THE ATMOSPHERE.

One of the most insidious and dangerous causes of lead pollution was Midgley's tetra-ethyl lead. When high-octane petrol is burned, minute particles of lead are released into the atmosphere. The highest concentrations of lead are found in the vicinity of busy roads and motorways as the particles settle in the immediate area. Even after tetra-ethyl lead had been phased out, there remained a risk to children playing around roads that were in use prior to the 1986 ban, because of the high concentrations of lead in the soil. The major European countries enacted their own bans in the 1990s, and China phased out tetra-ethyl lead in 2001.

Although the phaseout of tetra-ethyl lead and many other products containing lead in the United States in the 1980s has meant that the overall amount of the metal in the environment has markedly decreased, the after-effects of high lead use have continued to plague the United States and the developed world. According to economist Rick Nevin, lead exposure accounts for as much as 65 to 90 per cent of the variation of violent crime in the United States and another nine countries. Nevin's theory claims that the high crime rate of the 1980s and its subsequent fall in the 1990s was caused by differential rates of lead exposure in children after and before the ban. He also found a correlation between high crime rates and urban neighbourhoods built next to major freeways.

His research outside the United States bears out his claims. In the United Kingdom, for example, leaded petrol was phased out a decade later than in the United States, and crime rates continued to rise there when they were beginning to fall in the United States. He has found similar correlations between violent crime rates and lead pollution in Mexico and Latin America.

Nevin's theory has since found further scientific backing. A 2002 study of adolescents arrested in Pittsburgh, Pennsylvania, found that they had blood-lead levels four times higher than a control sample of high-school teenagers. According to the researchers, lead poisoning decreases self-control and makes an individual more likely to break the law. Ellen Silbergeld, professor of environmental health sciences at Johns Hopkins University, has concluded: 'There is a strong literature on lead and sociopathic behaviour among adolescents and young adults with a previous history of lead exposure.'

FAILING

Never got off the drawing board

Didn't work in practice

Killed its inventor

A commercial failure

Unforeseen consequences

Was used for evil ends

A success born of failure

CHITTY-CHITTY BANG CRASH: THE FLYING CAR

Main Culprits: Henry Ford (1863–1947), Waldo Waterman (1894–1976) and others

Motivation: Commercial venture

Damage Done: None at all – but think of the congestion, parking and pollution issues! And an airborne shunt would involve a lot more than some panel beating…

'When most people think about a flying car, they think about the convenience of door-to-door travel which such a machine might bring to their lives, and about the $400 hangar rent they could save by keeping it in their garage. Above and beyond these not insignificant speculations, what is not readily apparent is the potential of the flying car to bring about a revolution in private aviation, which could, in time, make flying as useful and as common as driving is today – that is, to finally bring about the old NACA dream of an airplane (flying car) in every garage.'

Advertising copy for the Volante Flying Car, 2008

In earlier articles we looked at human- and rocket-powered flight (see pp. 10–15 and pp. 60–3) and personal ground transportation (see steam car, pp. 94–8), and internal combustion engines (see pp. 108–12). Put these together and you get every boy and his dad's dream invention: the flying car. There is no shortage of fictional flying automobiles, from the vintage *Chitty-Chitty Bang Bang* to the futuristic cab flown by Korben Dallas (played by Bruce Willis) in *The Fifth Element*, but the real thing is fairly thin on the ground, or, in this case, in the air. In the early heady days of the airplane and the automobile, combining the two seemed to be an obvious idea, and several pioneers of flight and driving applied themselves to the problem, but with little or no success. However, with road congestion worsening by the year, could it be that the flying car's moment has finally arrived?

IN THE AIR AND ON THE ROAD

Until the invention of the internal combustion engine, the only way to get a craft airborne and keep it there for any length of time was to make it lighter than air. In the nineteenth century, balloons, dirigibles and airships (see pp. 131–5) ruled the heavens. But once a small, powerful engine had been invented, the sky was literally the limit. Everyone knows about the Wright brothers' flight in 1903, but few might be familiar with the names of two other aviation pioneers, Glenn Curtiss (1878–1930) and Waldo Waterman (1894–1976). Curtiss was the first man to design a flying car in 1917, the 'Autoplane', which was exhibited but never flew. His vehicle was a boxy car with a biplane wing fixed to the roof and a large push-propeller engine at the rear.

The tail-less 'Autoplane' inspired Waterman to create his own designs, which he worked on through the 1930s and '40s. He launched several models, including the 'Whatsit' and the more successful 'Aerobile'. The Aerobile was also tail-less and powered from the rear but was equipped with a monoplane wing. It had a wingspan of 38 ft (11 m) and a length of just over 20 ft (6 m). It was powered by a Studebaker engine, with a top air speed of 112 mph (180 km/h) and top ground speed of 56 mph (90 km/h). Two Aerobiles flew from California to Ohio, but the machine was never put into commercial production because of a combination of technological problems and regulations limiting its use.

In 1926, Henry Ford (1863–1947), the creator of the first mass-produced automobile, the Model T Ford, interested himself in producing the airplane equivalent, which he dubbed the 'Sky Flivver' ('Flivver' being the nickname of the Model T). His design, however, was not for a flying car but a small single-seater airplane for domestic use. The project was abandoned in 1928 when the pilot was killed in a test flight.

Three decades later Ford Motors tried again with the 1958 Ford Volante concept car. The Volante was an early version of the VTOL (vertical take off and landing) aircraft. It was powered by three large turbo fans, two to the rear and one to the front, that would lift the car vertically from a parking bay and into the air. The car never got beyond the model stage, but according to Ford, the concept was technically feasible. The proposed markets for the Volante included the police, armed forces and rescue services, as well as luxury personal transportation. A celebrity in a Volante would certainly give the paparazzi a headache. The project failed, however, at the regulatory level. Air traffic control in the 1950s was not computerised, and the potential increase in congestion would have overwhelmed the systems of the day with disastrous consequences.

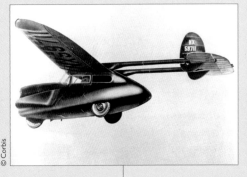

© Corbis

FLYING CAR
An experimental 'flying car', a fusion of auto and airplane, with an engine capable of 130 horsepower. Photographed 1946.

The 1940s to 1970s were something of a golden age for flying car inventors. An early technical success was the 'Aerocar', which first flew in 1949. The Aerocar was the brainchild of inventor Moulton Taylor. Although Taylor's design is reminiscent of Waterman's Aerobile, being a car body with wings attached, it had foldable wings so that it could be converted from an airplane into a car in minutes. The same engine drove the front wheels through a three-speed manual transmission and the propeller shaft. The Aerocar carried two people, the pilot and one passenger. It was 21 ft 6 in (6.55 m) long, with a wingspan of 34 ft (10.36 m) and a height of 7 ft 2 in (2.18 m). It had a top speed of 112 mph (172 km/h) and a range of 300 miles (480 km), though a later model increased this to 135 mph (216 km/h) with a range of 350 miles (560 km). On the road the Aerocar has a top speed of 60 mph (96 km/h). Although the Aerocar was certified in 1956, Taylor could not find enough buyers to make the project commercially viable. In the end, only six Aerocars were built, one of which was still flying at the time of writing.

The AVE Mizar, designed and built by Henry Smolinski between 1971 and '73, used an entirely different concept from the flying cars we have discussed so far. The Mizar was a hybrid vehicle consisting of the wing and tail sections of a Cessna Skymaster and a Ford Pinto. The passenger compartment and engine were removed from the plane, leaving a frame that was attached to the car by struts. The Mizar was powered with two engines, one under the hood for surface travel, and a second pusher-prop engine attached to the airframe. In flying mode, the aircraft section was bolted on, and the craft used both engines to take off. On landing the car braking system engaged, allowing the Mizar to come to rest in as little as 525 ft (160 m). For road use, the frame carrying the wings, tail and aircraft engine was detached and stored at the airfield, and the car drove off looking more or less like any other Ford Pinto, apart from the unusual roofline and an airplane cockpit instrument panel below the dashboard. The Mizar could carry four: a pilot and three passengers. It was 28 ft 6 in (8.6 m) long and 8 ft 6 in (2.6 m) high, with a wingspan of 38 ft (11.58 m). It cruised at an average speed of 130 mph (209 km/h), a maximum speed of 170 mph (273 km/h) and a cruising height of 16,000 ft (4,876 m).

Once certification had been obtained from the Federal Aviation Administration, commercial production was scheduled to begin in 1974, with several models ranging in price from US $18,300 to $29,000. Through 1973, Smolinski and his test pilot, Harold Blake, took the Mizar on several trial flights. On September 11, 1973, during a flight at Oxnard, California, the car and airframe parted company, and the now wingless Ford Pinto plummeted Earthward. Both Smolinski and Blake were killed, bringing an end to the Mizar project.

The failure of the Aerocar and the tragic crash of the Mizar did not put an end to the dream of a workable flying car. Today about a dozen inventors are taking the concept forward using different technologies ranging from the helicopter rotor to a roadable version of the VTOL aircraft. According to the London *Times* (January 11, 2009), the latest addition to the flying car family is the 'Terrafugia Transition', scheduled

DURING A FLIGHT AT OXNARD, CALIFORNIA, THE CAR AND AIRFRAME PARTED COMPANY, AND THE NOW WINGLESS FORD PINTO PLUMMETED EARTHWARD.

to go on sale in the United States in 2011. The two-seater Transition is a modern take on the Aerocar concept, with a push-propeller and folding wings, which Terrafugia claims is small enough to fit into the average North American garage. Improved materials and technology mean that it is far superior to its 1949 predecessor: the wings retract automatically in 15 seconds, and in aircraft mode it has a 500-mile (800-km) range at a cruising speed of 115 mph (185 km/h), all on a single tank of standard unleaded petrol. The Transition will retail at a hefty US $200,000 (£130,000), which, given the economic situation at the time of writing, might present an insuperable difficulty.

GROUNDED Even if inventors and manufacturers manage to overcome the many technological difficulties in adapting a vehicle to both road and flight, and solve the pricing problem, the flying car has always faced formidable regulatory issues. The FAA has classed all existing flying cars as aircraft, meaning that you need a pilot's licence to fly one, and that you can only take off and land from an airfield, which militates against their use as a convenient commuter vehicle, unless you happen to have an airstrip in your backyard. Additionally there are a host of safety issues to deal with: drunk drivers, drivers without a pilot's licence, accidents in built-up areas, and air traffic control problems, to name the most obvious ones. Last but not least, the flying car would be the ideal weapon for the suicide bomber and terrorist. Combined with environmental concerns, these issues mean that the personal flying car may remain grounded forever.

MONKEYING AROUND: QUACK REJUVENATION AND POTENCY THERAPIES

FAILING

Never got off the drawing board

Didn't work in practice

Killed its inventor

A commercial failure

Unforeseen consequences

Was used for evil ends

A success born of failure

Main Culprits: Serge Voronoff (1866–1951) and John Brinkley (1885–1942)

Motivation: Scientific inquiry and greed

Damage Done: Exploited the gullible and desperate; caused unnecessary illness

'The rationale for testicular transplantation superimposed modern interpretations on the older Victorian notion of the conservation of (seminal) energy.'

Elizabeth Watkins, *The Estrogen Elixir,* 2007

From antiquity to the early modern age, physicians associated male potency, youth and vigour with the retention of semen, a substance that was believed to have pseudo-magical properties. This lead to the prohibition, vigorously enforced by Victorian doctors, parents and educators, against masturbation, and to the conviction that a boy or man who 'spilled his seed' was not only sinful in the eyes of the Lord but also putting himself at risk of severe mental and physical debilitation. This was based on a misconception of how the hormone testosterone regulates sexual function and secondary male characteristics.

TESTOSTERONE

The male testes secrete testosterone, which plays a vital role in the development of male sexual characteristics during puberty. Unless there is a medical problem preventing the testes from producing testosterone, the body produces as much as it needs. It is true that testosterone production drops off with age, which several doctors claim leads to the 'andropause', or the male menopause, for which they prescribe a course of hormone replacement therapy (HRT). However, this is a controversial view, and many researchers point to the risks of taking HRT. Men taking high doses of testosterone experience the same negative side effects suffered by bodybuilders taking anabolic steroids to build muscle mass. These include acne, pattern baldness, infertility, gynecomastia (development of breast tissue in men) and the risk of accelerating an existing prostate cancer. In the early part of the twentieth century, however, before testosterone was discovered, doctors and quacks developed a much more hands-on approach to treating male sexual dysfunction: by grafting the testes of animals into human patients.

In the mid-nineteenth century, doctors began to understand the role of hormones in the regulation of metabolism, though they did not discover the chemical substances we call hormones until much later. One early pioneer of hormone research was the Mauritian-American doctor Charles-Édouard Brown-Séquard (1817–94). Having understood the role of the adrenal gland, he theorised that the testes produced a substance that could restore vitality and potency in aging men. As he himself aged, he injected himself with his own 'elixir', a preparation made from the testes of guinea pigs and dogs, hoping to prolong his life. Brown-Séquard's work inspired another doctor, the Russian-born

Serge Voronoff (1866–1951), who trained and practised as a surgeon in Paris. He first experimented with Brown-Séquard's elixir but the injections did not retard aging or improve potency. Voronoff decided to try out a much more radical procedure: *xenotransplantation* – the transplantation of animal tissue into human patients.

Voronoff's early experiments involved the transplantation of the thyroid glands of chimpanzees into patients suffering from thyroid deficiency. Thinking incorrectly that the testes were a type of gland like the thyroid, and produced the same type of regulatory hormone, he began to transplant the testicles of criminals who had been guillotined into the scrotums of wealthy patients, claiming that they would be rejuvenated and made more sexually potent. However, the good doctor soon ran out of spare human testicles, and he had to turn to those of our closest relatives, the great apes and monkeys. He made his first 'monkey-gland' xenotransplant in 1920, implanting not a whole testicle but thin slices of chimpanzee and baboon testicular tissue, which he believed would fuse with the human testes. At first his work was hailed as a breakthrough by the medical profession and patients rushed to be treated. Other surgeons took up the technique and by the 1930s, thousands of wealthy elderly men were walking around with slivers of monkeys' testicles grafted to their own family jewels.

VORONOFF'S EARLY EXPERIMENTS INVOLVED THE TRANSPLANTATION OF THE THYROID GLANDS OF CHIMPANZEES INTO PATIENTS SUFFERING FROM THYROID DEFICIENCY.

Many of his patients reported renewed vigour and sexual potency, making Voronoff a very wealthy man. However, as the science of endocrinology developed, his work was discredited and he became the subject of ridicule. We now know from work with human-to-human transplantation that rejection is a major problem that has to be suppressed with powerful immunosuppressant drugs. Hence Voronoff's animal grafts were instantly rejected and destroyed by the body's immune system, and any beneficial effects reported were due to the placebo effect.

Doctors such as Brown-Séquard and Voronoff were misguided, but they were not dishonest. They believed in their theories and were happy to experiment on themselves. Such was not the case of the American quack medical practitioner John Brinkley (1885–1942). Brinkley's bizarre career began in 1918. Claiming to be inspired by Voronoff's work, he implanted the testes of goats and sheep into male patients, claiming that the procedure would restore their potency and fertility. The operation had none of the refinement of Voronoff's procedures. Brinkley merely inserted the animal gonad into the scrotum, without any attempt to attach it to the body. The results were the same, however; the patient's immune system would attack and destroy the foreign body and any improvement was due to the placebo effect. In all, Brinkley is thought to have carried out this procedure on 16,000 people, many of whom suffered from post-operative infections. At his trial it was claimed that as many as 43 patients died as a direct result of Brinkley's ministrations.

MAD SCIENCE

Brinkley promoted his medical practice through his own radio station KFKB (Kansas First, Kansas Best). In 1930 both his medical and radio broadcast licences were revoked by the state of Kansas. In response, he stood for election as governor of Kansas, winning close to 30 per cent of the vote in 1930 and 1932. Having failed in his political ambitions, he opened a new radio station just across the Texas border in Ciudad Acuña, Mexico. He used the new radio station, which could be heard as far north as Canada, to restart his dubious medical practice, offering Mercurochrome injections to restore male potency. Brinkley's support for Nazi Germany during World War II led the US government to pass the Brinkley Act to regulate cross-border broadcasting from Mexico. In 1941, he was finally put out of business. His radio station was seized by the Mexican authorities and he was declared bankrupt. In 1942, the bogus doctor himself succumbed to illness, dying penniless of coronary heart disease.

FLYING HIGH: THE DISCOVERY OF LSD

FAILING

Never got off the drawing board

Didn't work in practice

Killed its inventor

A commercial failure

Unforeseen consequences

Was used for evil ends

A success born of failure

Main Culprit: Albert Hofmann (1906–2008)

Motivation: Scientific inquiry

Damage Done: Mental illness, injury and death of recreational users; abuse of medical experiments by the secret services of the United States, Canada and the United Kingdom

'...what one commonly takes as "the reality", including the reality of one's own individual person, by no means signifies something fixed, but rather something that is ambiguous – that there is not only one, but that there are many realities, each comprising also a different consciousness of the ego.'

Albert Hofmann, *LSD: My Problem Child*, 1980

Certain illicit drugs are associated with specific historical and cultural moments. The opiates laudanum and morphine became the drugs of choice of the nineteenth-century intellectuals, artists and writers. The illicit drug of our own time is another opiate, cocaine, in its crystalline form among the rich and famous, and as 'crack cocaine' on the inner-city streets. The drug that epitomises the counter-culture of the 1960s, however, was a totally synthetic creation, LSD (lysergic acid diethylamide), also known as 'acid'.

One might expect that LSD was the invention of a Californian counter-culture guru turned chemist, but the drug was actually first made in the staid, sleepy town of Basel, Switzerland, in 1938. It took some time for the full psychological effects of LSD to be known. Before its recreational use, LSD was a prescription drug used by the medical and psychiatric professions, and in the 1960s it was the subject of sinister experiments by the CIA.

'BICYCLE DAY' In 1938, the Swiss chemist Albert Hofmann (1906–2008) was working on a family of compounds looking for new drugs that would act as respiratory and circulatory stimulants when he stumbled on LSD. When animal tests showed little or no effect, apart from the subjects appearing 'restless', Hofmann discontinued the study. It was only five years later, in 1943, when he accidentally exposed himself to a minute dose of LSD through his skin, that he began to understand the properties of the chemical he had discovered. He wrote in his journal that on his way home from the laboratory, he had become dizzy and was affected by a 'remarkable restlessness, combined with a slight dizziness'. Once home, he lay down on his bed and sank into a pleasant intoxicated condition, which was characterised by 'an extremely stimulated imagination'. He described a dream-like state in which with his eyes closed he could see streams of 'fantastic pictures, extraordinary shapes with intense, kaleidoscopic play of colours'. The first 'acid trip' in history lasted for about two hours.

Three days later, he self-administered a 250-microgram dose of LSD, on what he would later refer to as the 'Bicycle Day'. Soon after taking the drug, Hofmann began to feel disorientated and had trouble speaking. He asked a colleague to take him home and both men rode back to Hofmann's house by bicycle. The ride on Bicycle Day was a classic acid trip. Everything around him was subject to distortion and strange motion. Although he was cycling at some speed, he felt he was stationary. Fortunately, he made it back home without incident. He was convinced that he was suffering from some form of poisoning, and he drank milk to try to relieve the symptoms. He then called his doctor, who could find nothing physically amiss with him apart from one of the classic physical symptoms of drug use, extremely dilated pupils.

In contrast to his benign initial experience of LSD, his second exposure to the drug was fast becoming the first 'bad trip'. Hofmann became delusional and was afraid that he was going mad. For the next few hours, he thought he was being possessed by a demon, that his next-door neighbour was a witch, and that his own furniture was trying to attack him. His doctor prescribed bed rest, and as soon as Hofmann lay down, he began to feel better. The paranoid delusions subsided and he once again experienced the display of psychedelic patterns when he closed his eyes. He described fantastic coloured images, as well as experiences of synaesthesia, when a stimulus on one sense causes a response in another sense, such as 'seeing a sound' or 'hearing a colour'. When Hofmann awoke the next day, he was tired but also felt clearheaded and invigorated. He wrote that his senses were 'vibrating in a condition of highest sensitivity, which then persisted for the entire day'.

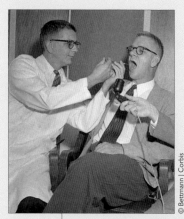

From 1948 until the mid-1960s, LSD, along with other hallucinogenic substances, was the subject of considerable scientific and medical research. The drug was used in psychiatric and psychotherapeutic settings for a range of mental disorders, including schizophrenia, alcoholism, sexual dysfunction and drug dependence. A trial in the 1950s reported that half of a group of alcoholics who had failed to quit by using more conventional methods, had not had a drink a year after treatment with LSD. In the United Kingdom, a specialist LSD unit was set up in 1958, and treatment continued there until 1972.

LSD EXPERIMENT
Dr Harry L. Williams (left) administers LSD 25 to Dr Carl Pfeiffer, to produce effects similar to those experienced by schizophrenics. The microphone shown was used to interview Dr Pfeiffer during the experiment.

One of the strongest advocates for the medical and recreational use of LSD was the Harvard psychology professor, Timothy Leary (1920–96). He was convinced of the therapeutic value of hallucinogenic drugs. He used the drugs on prisoners hoping that it would prevent them from reoffending. He reported a 90 per cent success rate, but it was later shown that he had skewed his data, and the actual improvement was in the order of 2 per cent. Leary's research and his promotion of the then legal drug among the Harvard student body led to his dismissal in 1963. Undaunted, Leary continued his experiments in counter-cultural living in Mexico and then in New York. However, by the mid-1960s, governments in the developed world were becoming alarmed by the growing rates of recreational hallucinogenic drug use, and they moved to stop the pharmaceutical production and later use of LSD. A United States Department of Agriculture report described the drug in the following terms:

> Although initial observations on the benefits of LSD were highly optimistic, empirical data developed subsequently proved less promising… Although the study of LSD and other hallucinogens increased the awareness of how chemicals could affect the mind, its use in psychotherapy largely has been debunked. It produces no aphrodisiac effects, does not increase creativity, has no lasting positive effect in treating alcoholics or criminals, … and does not generate immediate personality change. However, drug studies have confirmed that the powerful hallucinogenic effects of this drug can produce profound adverse reactions, such as acute panic reactions, psychotic crises and 'flashbacks', especially in users ill-equipped to deal with such trauma.

BANNED HIGHS

LSD became a controlled substance in the United States in 1966. Recreational use continued to increase, however, and high-profile users, such as musicians The Beatles, The Rolling Stones and Jimi Hendrix, admitted experimenting with the drug. LSD remained popular through the 1970s and '80s. Its popularity waned in the 1990s as it was not suited to the 'rave' culture generation, which favoured a new drug, MDMA, better known as ecstasy.

One of the more bizarre uses of LSD was its role in the CIA's classified MK-ULTRA project, which operated through the 1950s and '60s at the height of the Cold War. MK-ULTRA investigated the possibilities of mind-control techniques, through hypnosis, brainwashing and the use of hallucinogenic drugs, including LSD. A congressional inquiry into the project reported that the drug was given to agency and government employees, military personnel, doctors, prostitutes, mentally ill patients and members of the general public. The drug was administered with consent in certain cases, but in many others it was given without the subjects' knowledge or consent. The agency employed extremely unusual means of recruiting subjects, such as setting up brothels where the drug could be administered and the subjects observed through one-way mirrors. The experiments often involved psychological torture, and it is claimed that many of these involuntary guinea pigs committed suicide or ended up in psychiatric institutions. The agency eventually concluded that LSD's effects were too varied and uncontrollable to make it of any practical use as a mind-control drug, and the experiments with LSD were discontinued.

LSD has not killed as many people as heroin (see pp. 136–40), and is certainly much less dangerous physically than many of the other chemicals discussed in this book. The apologists of LSD saw it as a way to reconstruct consciousness by opening the human mind to new experiences, thus stimulating individual creativity and social transformation. However, like other psychoactive drugs, licit and illicit, it gives the illusion of expanded consciousness while locking the user into a narrow, limited, self-referential world cut off from reality.

FAILING

Never got off the drawing board

Didn't work in practice

Killed its inventor

A commercial failure

Unforeseen consequences

Was used for evil ends

A success born of failure

SILENT SPRING: DDT

Main Culprit: Paul Müller (1899–1965)

Motivation: Scientific inquiry

Damage Done: Collapse in bird populations; health problems in humans; worsening of insect-borne epidemic diseases

'After several years of DDT spray, the town is almost devoid of robins and starlings; chickadees have not been seen on my shelf for two years, and this year the cardinals are gone too … It is hard to explain to the children that the birds have been killed off, when they learned in school that a Federal law protects the birds from killing or capture. "Will they ever come back?" they ask, and I do not have the answer. The elms are still dying, and so are the birds. Is anything being done? Can anything be done? Can I do anything?'

A housewife from Hinsdale, Illinois, quoted in Rachel Carson's
***Silent Spring*, 1962**

In the late nineteenth century, when humans had finally got hold of the keys to the chemistry toy box, researchers began to discover and synthesise a huge variety of chemical compounds, often with little or no idea of what they would do once they were released into the environment. Such was the case of polythene (see pp. 141–5), which only came into its own in the mid-twentieth century. A similar pattern occurred with Dichloro Diphenyl Dichloromethane, the insecticide better known to us as DDT.

The German-Austrian chemist Othmar Zeidler (1859–1911) was the first to synthesise DDT in 1874, but its properties were not understood until 1939. There are differences between the two cases, however. Although polythene was in time recognised as being commercially significant and a useful new material, DDT was hailed as a miracle substance with the potential to save millions of human lives. Paul Müller (1899–1965), the Swiss chemist who discovered its insecticidal properties, was awarded the Nobel Prize for medicine in 1948 for 'his discovery of the high efficiency of DDT as a contact poison against several arthropods'. What was not realised at the time was that DDT, far from being a life-giving and life-enhancing substance, would turn out to be a silent, insidious killer.

BITING BACK

In order to grasp the full impact of DDT in the 1940s and '50s, we have to understand the role of insect carriers, or 'vectors', in the epidemic diseases malaria, yellow fever and typhus. The first two are transmitted to humans by the bites of different mosquitoes: the genus *Aedes* for yellow fever and the genus *Anopheles* for malaria; typhus is transmitted to human by fleas, lice and mites. Prior to the discovery of vaccines and effective drug treatments, these diseases caused millions of deaths worldwide. In the United States in the nineteenth century, for example, yellow fever epidemics killed 8,000 in New Orleans (1853), 3,000 in Norfolk, Virginia (1855), and 5,000 in Memphis (1878). During World War I (1914–18), typhus outbreaks caused three million deaths in Russia alone, and malaria, against which there is still no vaccine, was endemic in many parts of Europe and North America until the middle of the twentieth century. There are an estimated 250 million cases of malaria every year, resulting in the death of one million people, many of whom are babies and children.

Serious outbreaks of epidemic diseases tend to occur during and immediately after wars, when health and sanitation infrastructures are disrupted. Many millions died from insect-borne epidemic diseases in World War I, and the same would have probably happened in World War II (1939–45) had it not been for the fortuitous discovery of DDT in 1939.

The new insecticide was used extensively in several theatres of war to eliminate disease vectors. In Europe, DDT eradicated typhus from several areas where it had been endemic. And unlike in World War I, when thousands of combatants had died of the disease in the trenches, DDT delousing protected the armies of World War II. The principle victims of typhus were the inmates of German concentration and prisoner-of-war camps, who died in their thousands because of poor sanitary conditions and insufficient medical care. A campaign of aerial DDT spraying was particularly effective in the South Pacific, where malaria was a major problem.

© Bettmann | Corbis

SPRAYING DDT
A demonstration of DDT spraying by planes set to take part in a large-scale anti-malaria programme in postwar Europe.

After the war, DDT was made available for use in agriculture and the chemical became one of the most widely used insecticides, massively increasing the amount of DDT in the environment. In 1955, the World Health Organization (WHO) initiated a worldwide programme to eradicate malaria, by using DDT to wipe out the malarial host, the *Anopheles* mosquito.

In its first years, the programme was extremely successful. Malaria disappeared from Taiwan, the Caribbean, the Balkans, large parts of North Africa, northern Australia and most of the South Pacific. It also dramatically reduced mortality rates on the Indian subcontinent. However, the effects were short-lived, as resistance soon emerged in many insect populations as a consequence of the widespread agricultural use of DDT. In many regions, early victories against the disease were partially or completely reversed, and in some cases the incidence of malaria actually increased. In 1969, the WHO abandoned the programme altogether, turning to other means to combat the disease.

As early as the 1940s, scientists expressed concerns over possible environmental and health dangers associated with DDT. However, these early fears received little attention and it was not until 1957, when *The New York Times* reported an unsuccessful struggle to restrict DDT use in Nassau County, New York, that the issue came to the attention of the naturalist Rachel Carson. William Shawn, editor of *The New Yorker*, suggested that she write an article for the magazine, which developed into the book *Silent Spring* (1962). Carson argued that insecticides, including DDT, were poisoning both wildlife and the environment and were also endangering human health. The book highlighted the disappearance of many songbirds as one of the most notable effects of unrestricted agricultural use of DDT (see quote).

Silent Spring was an instant bestseller, and is now recognised as one of the forerunners of the environmental movement in the United States, as well as one of the best non-fiction books of modern times. DDT was a prime target of the growing anti-chemical and anti-pesticide lobby, and in 1967 a group of environmentalists founded the Environmental Defense Fund (EDF), whose main aim was the banning of DDT. The US Environmental Protection Agency (EPA) held seven months of hearings in 1971–2, at the end of which it announced the ban of most uses of DDT. The insecticide was subsequently banned for agricultural use worldwide under the Stockholm Convention in 2001, although it is still used to control insect disease vectors in several regions, including India and Sub-Saharan Africa, where it is sprayed inside houses or on mosquito nets.

DDT is particularly dangerous because it is quickly absorbed by the soil, where it has a half life of up to 30 years. When it leaches into an aquatic ecosystem, DDT is absorbed by organisms and the soil, or dissolves in the water itself. DDT breaks down into DDE and DDD, which are also highly persistent organic pollutants and have similar chemical and physical properties to DDT. These products together are known as 'total DDT'. DDT travels from the warmer regions of the world to the Polar regions by a process known as 'global distillation'. DDT, DDE and DDD accumulate in the food chain, with the top predators, such as birds of prey, having the highest concentrations and suffering the greatest ill effects.

DDT IS PARTICULARLY DANGEROUS BECAUSE IT IS QUICKLY ABSORBED BY THE SOIL, WHERE IT HAS A HALF LIFE OF UP TO 30 YEARS.

It is not only animals and birds that are at risk from DDT. Although it is classed as a 'moderately hazardous' substance by the WHO, which allows its use in spray form on the inside of houses and clothes to destroy insect disease vectors, there is evidence of health issues among people who have had long-term exposure to the chemical through working in its manufacture or its use in agriculture and insect eradication programmes. These include neuropsychological and psychiatric conditions and an increased incidence of non-allergic asthma.

There is some evidence, though none of it conclusive at the time of writing, linking DDT to increased risks of liver, testicular, prostate, breast and colorectal cancers. However, the link is thought strong enough for the US government to class DDT as a 'possible' human carcinogen. Finally, one study has shown that eating DDE-contaminated fish is linked to an increased incidence of diabetes.

Like the stories of so many drugs (see pp. 218–22), humankind's search for a cure to the world's ills can sometimes be so desperate that the side effects are ignored. Whether a blind eye has been turned to the potential for unpalatable consequences is open to debate in each individual case; however, it is clear that the pattern repeats itself too often to be simple misfortune.

DEATH SPRAYS: CFCs AND OZONE DEPLETION

Main Culprit: Thomas Midgley Jr (1889–1944)

Motivation: Scientific inquiry

Damage Done: Depletion of the ozone layer; increased risk of skin cancer; worsening of global warming; threat of human extinction

'The first significant loss of atmospheric ozone was observed over Antarctica in 1982. By the late 1980s ozone levels over certain Antarctic locations were down 60 per cent from 1955 readings during the spring season. In some areas ozone was almost totally absent from the air column between the altitudes of 15 and 20 km [approximately 9 and 12 miles]. The horizontal size of the so-called Antarctic 'ozone hole', a phenomenon of the spring season, became larger each year. By the early 1990s it was covering an area comparable to North America and extending outward to include parts of New Zealand, Australia, Argentina and Chile.'

Marvin Sooros, *The Endangered Atmosphere*, 1997

© Zagorskid | Dreamstime.com

All inventors have their off days. And some extremely able and successful scientists and engineers – such as da Vinci (see pp. 10–15 and 54–9), Brunel (see pp. 104–7), Tesla (see pp. 156–9), Curie (see pp. 146–9) and Sinclair (see pp. 244–6) – feature in this book for their one signal failure or disastrous invention. Thomas Midgley Jr (1889–1944) is unique in the annals of invention, however, for having discovered two chemical compounds that together have earned him the dubious distinction of being the single organism who has had the most impact on the atmosphere in the Earth's history. His many innovations included tetra-ethyl lead (see pp. 169–73), the toxic fuel additive that poisoned millions of humans, and the one discussed in this article, the family of chlorofluorocarbons (CFCs), which have put the very existence of life on Earth in peril.

COLD COMFORT Seven years after his work on tetra-ethyl lead, Midgley made another breakthrough, this time in the world of refrigeration. In 1928 he discovered the first of the family of chemicals known as the chlorofluorocarbons, which he called 'Freon' (CCl_3F) and which bonds three chlorine atoms and one fluorine atom to a carbon atom. Freon replaced the toxic ammonia and sulphur dioxide, and the explosive chloromethane, which had previously been used as coolants in both refrigerators and air-conditioning units.

The new compound was non-toxic, non-reactive and non-explosive. Midgley demonstrated its safety by inhaling a lungful and blowing it out onto a naked flame. He had made a similar safety demonstration with vaporised tetra-ethyl lead, worsening his own lead poisoning but in the case of Freon, the gas was as advertised and was indeed harmless to humans, at least when inhaled.

In the postwar period, CFCs and their extended family of related compounds had a wide range of industrial and consumer applications, as flame-retarding and fire-extinguishing agents, plastic foam expanders, refrigerants, propellants for aerosol sprays, cleaning agents for electronics and solvents. Unlike the main greenhouse gas carbon dioxide, which has many natural sources, the CFCs released into the environment have all been manufactured by humans since the 1920s, the bulk being produced in the 1960s.

In the 1970s, scientists studying the upper atmosphere noticed that there was something strange going on in the stratosphere, with the Earth's protective 'ozone layer'. In 1973, James Lovelock – most notable for his 'Gaia hypothesis' – took the first measurements of CFCs at both poles, noting their presence in all the samples he collected and analysed. Lovelock, however, did not realise what the CFCs were doing to the ozone. It wasn't until 1974 that two other climate scientists, Sherwood Rowland and Mario Molina, understood the havoc that CFCs were wreaking in the Earth's fragile upper atmosphere.

One of the properties that made CFCs so attractive for commercial application was their inertness: they don't burn, explode or break down into their constituent atoms at ground level. They are the chemical equivalent of a brick. Near the Earth's surface they cause little or no damage. Unfortunately, it is this inertness that also makes them extremely durable. CFCs have lifespans that are measured in decades, which gives them more than enough time to travel to the outer reaches of the earth's atmosphere. Once in the stratosphere, CFCs are subject to the full force of the sun's ultraviolet (UV) radiation, which is strong enough to break the bond holding the chlorine atoms. Unlike the CFCs, chlorine (Cl) itself is not an inert element; it is an extremely reactive 'free radical' – imagine something like a pinball launched into the machine.

Here comes the fun science bit. When the released chlorine (Cl) atom collides with an ozone molecule (O_3), it breaks the ozone down into chlorine monoxide (ClO·) and oxygen (O_2). The chemical shorthand for this is: $Cl· + O_3 \Rightarrow ClO· + O_2$. Chlorine monoxide is also a highly reactive substance, and it can also break down ozone in the following way: $ClO· + O_3 \Rightarrow Cl· + 2O_2$. A single rogue atom of chlorine can destroy as many as 100,000 ozone molecules.

You might think that doesn't sound too bad. When ozone breaks down, it creates oxygen, and oxygen is not a poison or dangerous to the planet. After all, without it, there would be no life here. This is true. However, the problem is not the creation of oxygen but rather the depletion of ozone. Ozone loss consists of two distinct processes: the decline of the total amount of ozone within the stratosphere, thought to amount to between 3 and 6 per cent every decade since the 1970s, and the much

GOING
STRATOSPHERIC

more dramatic effect at the South Pole, known as the 'ozone hole', which occurs every spring, when reductions can be as high as 70 per cent of pre-ozone hole values.

The role of ozone in the atmosphere is to absorb UVB radiation produced by the sun. The more ozone, the more UVB is absorbed; the less ozone, the more gets through to the Earth's surface. At present the loss of ozone at most latitudes is not thought to be significant, but at certain times of the year, the Antarctic ozone hole extends over parts of New Zealand, Australia and South America, raising fears of major health problems in these areas. UVB is a contributory factor in the development of the skin cancers carcinoma and malignant melanoma.

© Gerry Penny | EPA | Corbis

DISPOSING OF CFCS
Old refrigerators being cleansed of the offending CFC gases before disposal, in Lewes, England.

While the former is not usually life-threatening, the latter has a mortality rate of 15 to 20 per cent. One study showed that a 10 per cent increase in UVB radiation was associated with a 19 per cent increase in malignant melanomas for men and 16 per cent for women. Increases in UVB would also lead to increases in tropospheric or ground-level ozone, which is also known to be a health risk because of its oxidant properties.

The plant and animal kingdoms are also very susceptible to UVB. Several important crop species, including rice, depend on a type of bacteria to fix nitrogen in its roots. If surface UVB increased significantly, these bacteria would be destroyed and the crops that depend on them would also fail, leading to mass starvation on a planetary scale. Oceanic plankton, which forms the base of the marine food chain, is also very susceptible to the effect of UV radiation.

A final effect of ozone depletion is the worsening of global warming, though this effect is complicated by the fact that ozone depletion produces both global warming and global cooling. Ironically, ozone loss may be slowing down the warming of the atmosphere. On the other hand, CFCs are also significant greenhouse gases.

The United States and several EU countries banned the use of CFCs in aerosols in 1978. When scientists observed a dramatic seasonal depletion of the ozone layer over the South Pole in 1985, the discovery resulted in the convening of an international conference in Canada in 1987. The resulting Montreal Protocol called for drastic reductions in the production and use of CFCs. In 1990, the London Conference voted to strengthen the protocol and called for the complete elimination of CFCs by 2000 in the developed world, and by 2010 from the rest of the planet.

The crucial role of Sherwood Rowland and Mario Molina, alongside that of Paul Crutzen, was recognised with the award of the Nobel Prize for Chemistry in 1995. The citation read, in part:

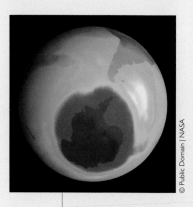

© Public Domain | NASA

> The thin ozone layer has proved to be an Achilles heel that may be seriously injured by apparently moderate changes in the composition of the atmosphere. By explaining the chemical mechanisms that affect the thickness of the ozone layer, the three researchers have contributed to our salvation from a global environmental problem that could have catastrophic consequences.

THE ANTARCTIC OZONE HOLE
An image created from satellite data captured on September 17, 2001, shows a hole roughly the size of North America. Amazingly, this represented a 'levelling-off of the hole size' prior to a predicted slow recovery.

The measures taken have led to major reductions in the emissions of CFCs, and the concentrations of the most damaging chemicals have been declining. By 2015, the Antarctic ozone hole is slated to shrink to 386,000 sq miles (1,000,000 sq km). However, we are not out of danger yet. Even if all goes according to plan, complete recovery of the Antarctic ozone layer will not occur until 2050 or later. Researchers estimate that a detectable increase in stratospheric ozone will not occur until around 2025, with ozone levels recovering to 1980 levels by around 2070–75.

FAILING

Never got off the drawing board

Didn't work in practice

Killed its inventor

A commercial failure

Unforeseen consequences

Was used for evil ends

A success born of failure

BARKING MAD: ANIMALS AS WEAPONS OF WAR

Main Culprits: The military

Motivation: Power and glory

Damage Done: Death and injury to millions of animals

'Animals in War. This monument is dedicated to all the animals that served and died alongside British and Allied forces in wars and campaigns throughout time. They had no choice.'

Inscription on Animals in War memorial by David Backhouse in London, unveiled in 2004

Before the mechanisation of warfare, animals served a variety of functions in the military. Horses, donkeys, mules, camels and elephants were used for transportation to and on the battlefield; pigeons, for communications; and dogs as combatants, guides, mascots and rescuers. Both live and dead animals have also been used as offensive weapons. The corpses of animals were sent into besieged towns to infect the population (see pp. 79–84), and one ancient Chinese general released burning live monkeys covered in straw and dipped in oil to start fires in the enemy camp.

With the advent of the internal combustion engine (see pp. 108–12), chemical incendiaries, aerial bombardment and radio communications, you might assume that that would be the end of the usefulness of animals in warfare. However, that would be to underestimate human ingenuity and human cruelty. Since World War II (1939–45) animals have been given new roles in war as living weapons.

During World War II, the Russians faced the mechanical might of the German army with limited technological resources of their own. The success of the invading Germans was largely due to the speed of their formidable columns of panzer tanks and armoured cars, which they used in their *blitzkrieg* ('lighting war') offensives in both France in 1939 and Russia in 1941. To stop these literally in their tracks, the Soviet Red Army engineers, who were critically short of conventional anti-tank mines and weaponry, devised a living tank-killer; these were known to the Germans as *hundminen*, 'dog mines', or anti-tank dogs.

The idea was as ingeniously simple as it was cruel – as a lot of really bad ideas are. The most vulnerable part of an armoured vehicle is its underside. So the Russians trained dogs to search for food under armoured vehicles and tanks. In preparation for battle, the dogs would be starved and fitted with a harness carrying an explosive charge on their backs. A simple trigger mechanism would detonate the explosive when the dog crawled under an enemy vehicle.

As the German armoured columns were sighted, the dogs would be released into their path. In theory, they should have homed in on the tanks, but in practice, in the thick of battle, the dogs often became confused or spooked. As they had been trained to find their food under

Russian tanks, they sometimes ran back toward their own lines, and blew up their own tanks, or they were simply scared by the roar of the engines or the noise of battle and ran away, sometimes causing panic by running back toward the Soviet lines. Estimates of their effectiveness vary, but it's thought that they may have destroyed up to 300 enemy tanks before the Germans started to use flamethrowers to repel the canine attacks.

BAT BOMBS

While the Soviets were pitting man's best friend against Nazi armour, their allies the Americans were coming up with animal capers of their own. Their targets were the fire-prone Japanese cities, whose domestic buildings were constructed of flammable wood with paper instead of glass windows. The idea was to drop bomb casings containing hundreds of thousands of bats fitted with incendiary devices at dawn over Japan's major urban and industrial centres. The bombs' descent would be slowed by parachute, and the casing was designed to open at 5,000 ft (1,500 m) to release their flying passengers. The bats, sensing the approach of dawn, would immediately seek out roosts in the city below. The incendiary devices were timed to go off once the bats were asleep deep inside buildings, igniting fires that would be difficult to extinguish.

The bat bombs were successfully tested on US soil on a mocked-up Japanese city built in Utah. On one occasion, armed incendiary bats released by mistake roosted under a fuel tank, causing considerable damage to an airbase in New Mexico. Although the bat-bombs had been shown to be effective, the project was abandoned in 1944 because the atomic bomb (see pp. 203–08) was almost ready to be deployed. If the bat-bomb had been finished first and used to devastate Japan's cities instead of the A-bomb, it is interesting to speculate how the future of humanity might have changed.

Staying with the flying theme, another US wartime scheme, known as Project Orcon (short for 'organic control'), was the brainchild of Harvard psychology professor and animal behaviour expert, B.F. Skinner (1904–90). The US armed forces were seeking a dependable guidance system for their missiles but the technology available at the time was too bulky to fit inside a missile warhead. Instead of an electronic brain, Skinner proposed using pigeons to guide missiles to their targets. The scheme called for a compartment containing pigeons in the nosecone of the missile. Skinner claimed that he could condition the birds to peck at screens displaying the missile's target. The missile's attitude controls would be linked to the screens so that the pecking would keep the missiles on target. The system would have used three pigeons as a fail-safe, and the missile would follow the majority decision of two birds. Although the military funded the project for over a decade, they cancelled Orcon as soon as non-organic systems became available.

© Bettmann | Corbis

MISSILE GUIDANCE
The pigeon subject of one of Harvard professor B.F. Skinner's psychological experiments. It must match a coloured light with a corresponding coloured panel in order to receive food. Photographed June 12, 1950.

Perhaps the best-known military animal of recent times is the military dolphin. Since the 1960s there have been rumours of dolphins trained in both the United States and the Soviet Union to seek out and destroy enemy divers, ships and submarines. While the Russians probably abandoned their programme after the collapse of the USSR in 1991, the Americans are believed by many to use both dolphins and sea lions for military purposes.

The US government denies that their sea mammals are used in offensive operations against humans, and claims that their role is to locate and rescue lost divers, to identify underwater mines and find enemy divers and report their whereabouts. These denials cut little ice with critics, however, and looking at the past record of the military and civilian agencies for insane schemes, I am inclined to agree. One can only hope that the dolphins are as intelligent as they are claimed to be and will have nothing to do with operations that involve humans and their own destruction.

The military are not the only ones to have hit upon the idea of using animals as living weapons of war. There have been several instances of mules and donkeys being used as improvised explosive devices (IEDs) by insurgents in Israel, the occupied territories, and lately in Iraq. The first recorded instance was in 1982, during the Israeli invasion of southern Lebanon, when a mule was used to carry explosives toward Israeli troops. In 2003, a donkey exploded near an Israeli checkpoint in the West Bank town of Bethlehem. There were no casualties apart from the donkey itself. In 2004, an exploding donkey was directed toward a US military checkpoint in Ramadi, Iraq, but again the IED exploded out of range of its intended target.

The above examples all highlight the main problem of using animals as weapons. Although they cannot understand the actions that we involve them in, they do have a sense of their own mortality and a strong instinct for self-preservation. An animal carrying explosives is perhaps more likely to run back toward its handlers than it is to charge into the path of advancing columns of tanks or infantry.

'ANIMALS CLAIM NO NATION'

You do not have to be an animal lover to feel a profound sense of revulsion at the use of animals as weapons of war, be it by terrorists or by the military of nation states. Animals are the ultimate innocents. They do not share our values, beliefs or hatreds. I shall leave the final words of this article to animal rights campaigner Ingrid Newkirk: 'Animals claim no nation. They are in perpetual involuntary servitude to all humankind, and although they pose no threat and own no weapons, human beings always win in the undeclared war against them. For animals, there is no Geneva Convention and no peace treaty – just our mercy.'

DESTROYER OF WORLDS: THE A-BOMB

FAILING

Never got off the drawing board

Didn't work in practice

Killed its inventor

A commercial failure

Unforeseen consequences

Was used for evil ends

A success born of failure

Main Culprit: J. Robert Oppenheimer (1904–67)

Motivation: Fear that somebody else would get there first

Damage Done: Illness, injuries and deaths from nuclear tests and the two bombs dropped on Japan in 1945; expenditure of trillions of dollars on weapons that can never be used; potential extinction of all life on Earth

'Out of 76,327 buildings, over 50,000 are destroyed. Up to 125,000 people will die on the day or will die soon. The wind mixes their dust with the dirt and debris. Then it sends everything boiling upward in a tall purple-gray column. When the top of the dust spreads out, it looks like a strange, giant mushroom. The bottom of the mushroom cloud is a fiery red. All over the city fires spring up. They rise like flames from a bed of coals.'

Laurence Yep, *Hiroshima*, 1996

Although many animals kill and eat members of their own species, and sometimes their own mates and young, humans are unique in the animal kingdom for their ability to contemplate the wholesale liquidation of their entire species, and by extension the extinction of every other species on the planet. Not content with the killing potential of gunpowder (see pp. 38–43), high explosives (see pp. 118–22) and chemical and biological warfare (see pp. 54–9 and 79–84), humans have gone on to create the ultimate weapon – a weapon so dangerous that it cannot possibly be used.

FALLOUT　The spectre of nuclear destruction has haunted the world since the detonation of the first atomic bomb on July 16, 1945. Within a few weeks, the first and only A-bombs to be deployed against a civilian population in wartime were dropped on the Japanese cities of Hiroshima on August 6 and Nagasaki on August 9. The effects of the explosions continue to this day, affecting the health of the men and women exposed to radiation and of their children through genetic damage. The total number of people who have died directly or indirectly from the two bombs had reached 400,000 by 2007.

The Russians tested their first atomic bomb in 1949, and the world entered the aptly named era of MAD (mutually assured destruction). The low point of the Cold War came with the Cuban Missile Crisis of 1962, when the world came closest to a full-scale nuclear war between the United States and NATO on one side and the USSR and Warsaw Pact on the other. Parodied in the movie *Dr Strangelove or: How I Learned to Stop Worrying and Love the Bomb*, the possible effects of nuclear war included the extinction of most plant and animal life on the surface of the planet, from the bombs themselves; the radiation that would poison air, land and water; and the nuclear winter that the mass detonations would trigger. Having pulled back from the brink, the world appeared to be a safer place when the Soviet Union collapsed in 1991 and the United States was left as the only global superpower. Unfortunately, in the intervening period, the world has witnessed the proliferation of nuclear technology. In addition to Britain and France, countries possessing nuclear weapons now include China, Israel, India and Pakistan, while two so-called 'rogue states', North Korea and Iran, are said to be working on their own nuclear-weapons programmes.

The discovery that atoms are not solid objects – like minuscule ball bearings – but solar systems of even smaller elementary particles orbiting around a central nucleus was made in the wake of the discovery of radioactivity by Becquerel and Curie (see pp. 146–9) at the turn of the nineteenth century. By the 1930s, physicists, including Albert Einstein (1875–1955), theorised that prizing apart the neutrons and protons that made up the atomic nucleus would release vast amounts of energy. Einstein expressed this potential energy in his famous equation $e = mc^2$, in which e (energy) is equal to m (mass) multiplied by c (the speed of light = 299,792,458 m/sec) squared. To put this in a layperson's terms, this would mean that a gram of mass (about one-third of an ounce) would release the equivalent energy to 21.5 kilotons (21,500 tons) of TNT, or 21.5 billion calories of heat.

This process is known as nuclear fission. In practice the nuclei of only a few elements are capable of being split in a controlled way. It would be very disturbing if the nuclei of common elements, such as carbon, hydrogen, and oxygen, could undergo fission on Earth, and if an atomically unstable ham sandwich, to pick an example at random, could take out a medium-sized town. The fissile elements that we use for power generation and nuclear weapons are uranium (U) and plutonium (Pu). These are both naturally radioactive substances; in other words, they decay, emitting radiation in the process. However, with these two elements a controlled chain reaction can be triggered so that fission can be sustained to power a reactor or an explosion. The physics of nuclear weapons were well understood by the 1930s, when the Nazis came to power in Germany, and several leading physicists, including Einstein, wrote to President Franklin D. Roosevelt (1882–1945) in 1939, urging him to develop nuclear weapons before the Germans did.

In response to the very real threat of a Nazi A-bomb, the United States, United Kingdom and Canada set up the Manhattan Project (1942–45) with the aim of making the first nuclear weapons. The man in charge of the scientific arm of the project was the gifted professor J. Robert Oppenheimer (1904–67). Oppenheimer was a physics prodigy. He skipped straight from high school to postgraduate physics, being exempted from undergraduate work, and went on to obtain his doctorate at the age of 22. He was also an accomplished Sanskrit scholar and

A GRAM OF MASS (ABOUT ONE-THIRD OF AN OUNCE) WOULD RELEASE THE EQUIVALENT ENERGY TO 21.5 KILOTONS (25,100 TONS) OF TNT, OR 21.5 BILLION CALORIES OF HEAT.

could read the Indian classics in their original editions. In some ways Oppenheimer was an unlikely choice to head such a sensitive research project. He was known to be a communist sympathiser – and several historians claim he was an active member of the American Communist Party in the 1920s and '30s, when the FBI kept him under surveillance. However, during World War II, the Soviets and Americans were allies, and it was only after the end of the war, when the Cold War started, that he fell victim to the McCarthy witch-hunts.

The Manhattan Project had three main sites: a plutonium-production facility at Hanford, Washington; the uranium-enrichment facilities at Oak Ridge, Tennessee; and the nerve centre of the operation, the weapons research and design laboratory at Los Alamos, New Mexico. At its height the project employed 130,000 people and cost approximately US $2 billion (US $24 billion at 2008 prices). The project created two designs for the A-bombs. *Little Boy*, which was dropped on Hiroshima, was a uranium-235 bomb.

TRINITY SITE
A camera observation post facing the Trinity test site.

There was no prior test of this bomb before it was dropped on Japan, because the physics of setting a chain reaction in the uranium isotope was thought to be foolproof. The second bomb, *Fat Man*, was a much more technically complex plutonium-239 device. *Fat Man*, which would be dropped on Nagasaki, was tested on July 16, 1945 at the Trinity test site in New Mexico. Oppenheimer later wrote that when the bomb exploded, he remembered two verses of the Indian Sanskrit classic the *Bhagavad Gita*: 'If the radiance of a thousand suns were to burst at once into the sky, that would be like the splendour of the mighty one', and 'Now I am become Death, the destroyer of worlds'.

After the success of Trinity, President Harry S. Truman (1884–1972) authorised the use of atomic weapons against two Japanese cities to bring the war in the Pacific to an end. The uranium *Little Boy* was detonated 1,900 ft (600 m) over Hiroshima at 8.15am, on Monday, August 6, 1945. The bomb took less than a minute to descend to its detonation height, but because of strong winds, it drifted 800 ft (244 m) off target, exploding over a surgical hospital. It caused a blast equivalent to 13,000 tons of TNT, which represented a mere 1.38 per cent of the uranium fissioning. Even so, the radius of total destruction was one mile (1.6 km). Three days later, at 11.01am local time, the plutonium bomb *Fat Man* was dropped over Nagasaki, exploding 43 seconds later at an altitude of 1,540 ft (469 m). This time the blast was even larger, and generated a yield equivalent to 21,000 tons of TNT, and a total destruction radius of one mile (1.6 km). The casualties from the initial blasts were startling: 70,000 in Hiroshima and 40,000 in Nagasaki. By the end of 1945, twice that number had died from burns, injuries or radiation sickness.

Immediately after the war, there was talk of dismantling the only nuclear arsenal then in existence and ensuring that no A-bombs would ever be built again, and instead developing the peaceful uses of nuclear technology under the auspices of the United Nations. However, such idealism and faith in human nature was entirely misplaced. With the defeat of the Germans and the Japanese, new enemies were found, and an undeclared war began between the United States and Soviet Union that would continue for the next half-century. The Soviets immediately began to work on their own A-bomb, using a combination of homegrown talent and espionage to reproduce the achievements of the Manhattan Project. The first successful Russian test took place on August 29, 1949, at a test site in the then Soviet republic of Kazakhstan in Central Asia.

Any hope of putting the nuclear genie back into the bottle was now abandoned. The two superpowers squared off against one another, initiating the nuclear arms race. Other powers also rushed to develop their own nuclear capabilities, beginning with the British in 1952, followed by the French in 1960 and the Chinese in 1964. These five countries represented the first wave of nuclear states.

NO GOING BACK

The second wave of nuclear countries was headed by Israel, which probably tested a nuclear device in 1979, but still denies having a nuclear capability. India detonated its first bomb in 1974, and Pakistan followed suit in 1988, creating the most dangerous nuclear flashpoint after the Middle East, because of the continuing tension between these two neighbours.

However, far more worrying are the 'rogue' states of the third wave of nuclear proliferation, which include North Korea and Iran. Although both countries have diametrically opposed political ideologies – the former being fundamentalist Communist and the latter fundamentalist Islamist – they are totalitarian states with no democratic controls and with links to international terrorism, and they have repeatedly threatened the annihilation of their respective enemies: South Korea and Japan in the case of North Korea; Israel in the case of Iran.

HIROSHIMA AND NAGASAKI

The locations of Hiroshima and Nagasaki in Japan, where the world's first two nuclear attacks effectively signalled the end of World II War. *Little Boy* was dropped over Hiroshima on Monday, August 6, 1945, while *Fat Man* was dropped over Nagasaki three days later.

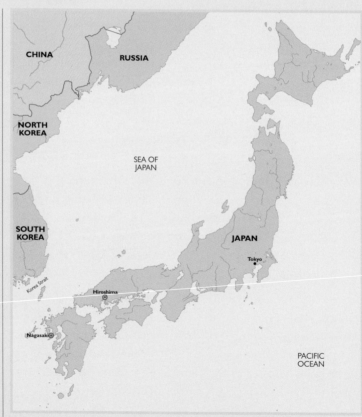

BEYOND BELIEF: L. RON HUBBARD'S DIANETICS

FAILING

Never got off the drawing board

Didn't work in practice

Killed its inventor

A commercial failure

Unforeseen consequences

Was used for evil ends

A success born of failure

Main Culprit: L. Ron Hubbard (1911–86)

Motivation: Greed

Damage Done: Swindled money out of millions; created a cult

'Serendipitously, the rise of Scientology coincided with the maturation of the generation of American Baby Boomers, whose self-absorption led to the development of what has been called the therapeutic culture. Hubbard's Dianetics, which has sold an estimated 20 million copies worldwide, was a masterful self-help book, and the length of time that it stayed on the bestseller list reflects the resonance of Scientology's appeal to Americans who reject the primacy of the medical – especially psychiatric – establishment.'

E V Gallagher, *Introduction to New and Alternative Religions in America,* 2006

Many of the medical quacks and pseudo-scientists of the nineteenth and early twentieth centuries used the major scientific breakthroughs of their day as covers for their scams. We've examined Ghadiali's Spectro-Chrome therapy (see pp. 165–8), various nostrums, cure-alls and elixirs (see pp. 127–30) and xenotransplantation (see pp. 179–82), each of which claimed some sort of scientific basis.

DIANETICS | L. Ron Hubbard (1911–86) tried a different and ultimately much more successful approach in eschewing magic potions and miracle machines that could be easily demonstrated to be ineffective, and creating his own science of the mind, 'Dianetics' (from the Greek 'through mind'), which is much more difficult to prove or disprove. Even respected psychotherapeutic and psychiatric theories of the mind and mental health are not backed up with the same kind of empirical evidence that can be obtained for a new drug therapy. Ultimately, Hubbard was to take Dianetics one step further by transforming it into a religion, 'Scientology', thus removing it from the domain of scientific evidence altogether. He is celebrated by his followers as a divinely inspired teacher and reviled by his opponents as a drug-crazed, money-grabbing charlatan.

Hubbard began his career as a writer of genre fiction, in particular, science fiction. He laid down the principle of Dianetics in an article published in *Astounding Science Fiction* magazine, which he later expanded in a book, *Dianetics: The Modern Science of Mental Health*, both published in 1950. In his writings, Hubbard explained that the mind consisted of the 'analytical' and the 'reactive' minds. The analytical mind is the conscious mind and the reactive mind is the unconscious. Traumatic experiences and memories from this and previous existences (including alien lives) are stored as negative 'engrams' in the reactive mind. Engrams are the cause of a range of mental and physical conditions, including arthritis, allergies, asthma, coronary heart disease, ulcers, migraine headaches and sexual deviations. The aim of Dianetics is to become 'clear' by the process of 'auditing' the reactive mind and eliminating the negative engrams.

The reader will recognise several common concepts from psychotherapy and psychoanalysis, such as the unconscious, repressed memories and trauma, and psychosomatic illness, which under different names

Hubbard claimed were his own original ideas. According to Hubbard, the benefits of going 'clear' were dramatic. In addition to being cured of many physical ailments, the subject would have no mental compulsions, repressions, psychoses or neuroses; he or she would enjoy a near-perfect memory, and his or her IQ would increase by as much as 50 points.

Despite the wealth of both Hubbard and the Church of Scientology, neither has tried to establish the truth of these claims through rigorous scientific testing. And because Dianetics was quickly dismissed as a pseudo-science by the medical establishment, it has seldom been the subject of independent investigation. However, New York University conducted two studies in the 1950s to test three of Hubbard's claims. The studies concluded that Dianetics did not significantly improve intellectual functioning, mathematical ability or the incidence of personality conflicts. In response, Hubbard's Dianetic Research Foundation presented anecdotal case histories of cures that were very short on solid evidence.

Dianetics, however, captured something in the popular imagination of 1950s America. *Dianetics: The Modern Science of Mental Health* became a bestseller, and Hubbard was able to establish his Dianetics Foundation in six major US cities with the proceeds from book sales.

MIND OVER MATTER

Within a few months, however, the fledgling Dianetics movement was in serious financial trouble. In 1951, the New Jersey Board of Medical Examiners forced the foundation to close for teaching medicine without a licence. In the following year, the foundation went bankrupt.

Having lost the rights to the name Dianetics, and to avoid any future entanglements with the medical authorities, Hubbard decided to rebrand Dianetics as 'Scientology', which he described as a 'practical religious philosophy'. Having the status of a religious organisation had several advantages in that his claims could no longer be evaluated on purely scientific grounds and that the church, like other religious bodies in the United States, would be exempt from federal and state taxes.

In 1953 Hubbard moved to the United Kingdom, from where he supervised the growing Church of Scientology from a Georgian manor house that had once belonged to the Maharajah of Jaipur. At the same time he developed Dianetics, adding a spiritual dimension that used auditing and a bio-feedback machine known as the 'E-meter' to reach a higher plane of human existence, which he called the 'Thetan'. Scientologist teachings also include beliefs in reincarnation and past lives and the existence of alien intelligences. In order to avoid the financial difficulties that had brought the Dianetics Foundation low, Hubbard instituted a system of lucrative fees and donations for courses, books and 'E-meters'.

It is easy to poke fun at Scientology's mix of psychotherapy, Eastern philosophy, self-help and sci-fi. However, that would fail to account for its tremendous popularity, and the adherence of several high-profile

celebrities. Many prophets are misunderstood and reviled in their time, and, in fact, since the time of Jesus, being rejected and despised seems to be a qualification for the job. Perhaps the measure of any idea or belief system is the man or woman who has created it. Abraham was a shepherd, Jesus a carpenter, and Muhammed a camel drover, yet the religions they founded have endured, in part because they embodied in their daily lives and actions the tenets of their faith.

L. RON HUBBARD
L. Ron Hubbard
photographed on January 10,
1982 by Michael Ochs.

However, according to many independent sources, Hubbard is also found wanting in this respect. Not only did he seriously misrepresent his childhood experiences and education, but there is also a high probability that he abused both alcohol and licit and illicit drugs during his lifetime. He earned a considerable personal fortune, which he spent on the worldly lifestyle of a celebrity rather than of a spiritual religious teacher. Finally, instead of being the mentally and physically perfect 'Thetan' that he claimed to be, Hubbard is known to have suffered from several common medical ailments. He died of a stroke at the comparatively young age of 74.

BIG BROTHER IS WATCHING YOU: TRAFFIC ENFORCEMENT AND SURVEILLANCE

FAILING

Never got off the drawing board

Didn't work in practice

Killed its inventor

A commercial failure

Unforeseen consequences

Was used for evil ends

A success born of failure

Main Culprits: Maurice Gatsonides (1911–98) and Walter Bruch (1908–90)

Motivation: Public safety and security

Damage Done: Loss of human rights

'The proponents of cameras argue that they save lives and eliminate any unfair subjectivity in traffic enforcement. Critics do not like the idea of being watched so extensively. They also worry about the temptation for governments and companies to raise money by putting cameras everywhere. Whether intentionally or by malfunction, there is also the risk that minor changes in the timing and calibration of the detectors could haul in a large number of additional fines from people who would otherwise not be committing traffic violations.'

Kevin Keenan, *Invasion of Privacy*, 2005

© Fintastique | Dreamstime.com

Not surprisingly, the surveillance camera in its various guises, and in particular the speed camera, often features on lists of the top ten worst inventions of all time – most likely voted there by drivers the world over. But should it be on the list at all? Traffic cameras have demonstrably reduced speeding and other traffic violations, thus reducing the toll of deaths and injuries; and the presence of surveillance cameras in our cities gives people a sense of increased personal safety.

1984 RETURNS

There are two distinct types of surveillance camera: those taking stills are used to catch speeding drivers in combination with a device that measures speed; and closed-circuit television (CCTV) cameras taking real-time video in stores and banks, and now increasingly on our streets. Together they provide a formidable surveillance arsenal for the state and private corporations, enabling them to keep track of where we go and what we do. In the famous novel *Nineteen Eighty-Four* published in 1949 by George Orwell (1903–50), the citizens of a dystopian future are under constant surveillance in their homes by the state – a prediction that, according to several human-rights activists, has already come true in the United Kingdom, the country with the most surveillance cameras per head of population on Earth.

When he invented the world's first speed camera, the Dutch rally driver and winner of the 1953 Monte Carlo Rally, Maurice 'Maus' Gatsonides (1911–98) had something very different in mind than catching speeding drivers. His aim, in fact, was the complete opposite: he was looking for a way to improve his cornering ability on the track and, therefore, for ways to increase his speed. He soon realised the potential of the 'Gatsometer' in law enforcement, however, and founded a company to market and sell his invention. The first 'Gatsos', as they are nicknamed in many European countries in his honour, used conventional film cameras to take pictures, requiring the time-consuming process of collecting the film and processing it. Since the 1990s, Gatsos have been fitted with digital photography and video technology, transmitting images directly to control centres for instantaneous processing. Maus's company, Gatsometer BV, remains the world leader in the manufacture of speed, radar and red-light cameras.

Gatsos now have a wide number of applications. In addition to speed cameras that use either radar technology to measure the speed of a vehicle, or a set of multiple cameras with number-plate recognition software that can check the average speed of a vehicle between two points, cameras are also used at red lights, prohibited turn signs, no-stop junctions and stop signs; in bus and car-pooling and high-occupancy lanes; at toll booths and level crossings; and to ensure that vehicles do not overtake in prohibited passing zones. In London, an extensive camera system fitted with number-plate recognition technology controls payment to the congestion zone that covers most of the centre of the city. The daily charge for the congestion zone is £10 (US $15) per day, and non-payment fines of £100 (US $150) ensure that the charge is paid. The latest development in the United Kingdom is the use of still cameras to monitor controlled parking zones.

Critics have attacked traffic-enforcement cameras on two grounds. The first is that they are a moneymaking scheme for national and local government. In areas such as congestion charging and parking enforcement, there is no safety issue, and drivers feel that they are subject to a form of 'stealth taxation'. The second criticism is that they infringe people's human rights by enabling government agencies to keep track of people as they travel. In addition, anti-camera campaigners have questioned whether speed cameras really save as many lives as it is claimed. Although it is true that at certain accident blackspots the introduction of cameras does decrease accidents, it is possible that they merely export speeding to areas where drivers know that cameras are not installed.

© Bettmann | Corbis

GEORGE ORWELL
Photographed here in the 1940s, George Orwell was born Eric Arthur Blair in 1903. He is best known for his anti-totalitarian works, *Animal Farm* and *Nineteen Eighty-Four*, the latter of which is seen by many as prescient of today's 'surveillance society'.

Although cameras have been on the increase on the world's roads, there is evidence that governments and law-enforcement agencies are reconsidering their use. The English town of Swindon was the first in the United Kingdom to discontinue the use of speed cameras in 2008. Friendswood, Texas, was the first US community to introduce speed cameras in 1986 and, due to a public outcry, one of the first to remove them. Compared to Europe, their adoption in the United States and Canada has always been far more contentious, and many individual communities, states and provinces have resisted their introduction.

The second arm of remote surveillance is closed-circuit television (CCTV). The German engineer Walter Bruch (1908–90) invented the first CCTV system for the Siemens Company during World War II for the launch site of the world's first modern rocket-powered missile, the V2 (see pp. 60–3) at Peenemünde, in northern Germany. The V2s were developed as a weapon of last resort to win the war for the Nazis. Bruch's system was used to monitor the launches from the underground bunker, much as cameras are used today in rocket and shuttle launches.

THE TOWN OF OLEAN, NEW YORK, WAS THE FIRST CITY IN THE UNITED STATES TO INSTALL CCTV ALONG ITS MAIN STREET IN AN EFFORT TO FIGHT CRIME.

The town of Olean, New York, was the first city in the United States to install CCTV along its main street in an effort to fight crime. The United Kingdom followed suit with a system in King's Lynn, Norfolk, in the early 1970s. CCTV soon became commonplace in banks and large stores, which fitted 'eye-in-the-sky' type cameras on their ceilings to deter thefts. In recent decades, the number of CCTV cameras in public spaces has increased considerably, especially in the United Kingdom.

The exact number of CCTV cameras in the United Kingdom is not known, but in 2002 campaigners Michael McCahill and Clive Norris estimated that there could be as many as half a million cameras in London alone, and a total of four million or more nationwide. This would mean that there was one camera for every 14 UK citizens, a number that is bound to have grown since the terrorist bombings on July 7, 2007 in London killed 52 and injured 700 commuters. Another group gives a figure of 1.5 million CCTV cameras located in UK city centres, railway stations, airports, major tourist attractions, shopping centres and financial institutions, but this figure does not include the growing number of private CCTV systems installed by store owners and, increasingly, homeowners. Although no exact statistics exist, the United States and Canada are thought to have far fewer CCTVs in public spaces than the United Kingdom, but their number is increasing rapidly.

Apart from the many 'Big Brother' houses that are televised for our amusement worldwide, we do not have surveillance cameras fitted as standard in our homes like the citizens of Orwell's disturbing novel, *Nineteen Eighty-Four*. Orwell can now be seen as prescient, though his predictions for a 'surveillance society' came true about a decade or so after the year given in his novel.

As with traffic-enforcement cameras, critics of CCTV have questioned its effectiveness in deterring and solving crimes. According to a 2008 report, only three per cent of crimes in the United Kingdom were solved with CCTV evidence. People are led to think that CCTV is far more effective because of a small number of high-profile cases, such as the 7/7 London bombings, when CCTV evidence was used to identify the bombers. However, for most crimes, the resources necessary to examine the hundreds of hours of recorded video are just not available. Often CCTV merely records the crime for posterity, as was the case in the Columbine High School massacre of 1999.

Privacy campaigners argue that CCTV presents a serious threat to civil liberties, because the face-recognition technology that is now being developed will allow government agencies to secretly monitor the activities of citizens, in order to impede their rights to free assembly and peaceful demonstration.

WHO IS WATCHING YOU?

Data obtained through CCTV is not currently controlled in the United States and can be kept indefinitely for use in a criminal investigation or legal proceedings years after the event. In addition to government agencies, large corporations also make covert recordings of their customers and employees, ostensibly on grounds of security, but which could be used for more sinister ends.

FAILING

Never got off the drawing board

Didn't work in practice

Killed its inventor

A commercial failure

Unforeseen consequences

Was used for evil ends

A success born of failure

MEDICINE OR POISON: IN SEARCH OF THE MAGIC BULLET

Main Culprits: Grünenthal Pharmaceuticals, among others

Motivation: Greed

Damage Done: Tens of thousands of deaths and children born with severe physical disabilities

'The striking defect that characterised many survivors was severely malformed or deficient extremities ... Belated experiments investigating thalidomide's toxicity showed that the drug could harm the embryos of mice, rabbits, chickens and monkeys. Sadly, these experimental results were too late to alert doctors to the risks of the drug in pregnant women. Between 1957 and 1962, as many as twelve thousand children were affected, mostly in Europe, Canada and Japan ... The thalidomide catastrophe was so shocking that it chilled the postwar euphoria of an imminent medical utopia.'

Morton Meyers, *Happy Accidents*, 2007

Contemporary society could well be described as a 'drug' culture, not just because of the proliferation of illicit narcotics in the postwar period, but also because of the growing use of prescription of medications for a wide range of physical and mental ailments. Although the benefits of many of these drugs outweigh their drawbacks, in a significant number of cases, drugs have caused serious injuries or death to patients. In some cases, as with cancer therapies and HIV anti-retroviral therapies, the potentially serious side effects are justified because they extend the lives of patients, but in others, when the benefits are much less clear-cut, no such risk–benefit trade-off can be made.

One example of such a trade-off is digitalis, the extract of the foxglove (*Digitalis purpurea*). Foxglove itself was known to the ancients as a potent source of poison, but it also had a variety of uses in folk medicine. In fact, the modern use of digitalis stems from an occasion when the physician William Withering (1741–99) heard of a gypsy remedy for heart problems that contained extract of foxglove. Withering experimented further and formulated a new medicine, although at the time its active ingredients were not understood.

KILL OR CURE

In a similar fashion, the products of the opium poppy (see pp. 136–40) have been used for both good and ill throughout history. The capacity of opium to dull pain has meant that while the illicit use of derivatives such as heroin can prove deadly, opium-derived opiates are among the most effective painkillers used today.

The secret of this dual nature lies in a drug's therapeutic index, the difference between the therapeutic dose and the toxic dose. In other words, as the sixteenth-century Swiss physician Paracelsus (1493–1541) put it: 'Only the dose determines that a thing is not a poison.'

Another example found within these pages is the use of arsenic in both medicine and murder (see pp. 44–8). Given that the proximity between that which can kill or cure has been known for so long, it perhaps seems surprising that so many mistakes continue to be made.

In fact, the German scientist Paul Ehrlich (1854–1915) sought to break this link. Inspired by his research in dyes, which he could use to mark specific cells such as bacteria, Ehrlich was intent on finding a way to not just mark but attack particular cells, leaving others untouched – a mechanism that he called a 'magic bullet'. Alongside his co-workers, Ehrlich synthesised hundreds of compounds, until he struck lucky with the arsenic-based compound NO. 606, which attacked the micro-organisms responsible for syphilis.

Unfortunately for Ehrlich, his solution was derided in some quarters for mitigating a disease that some considered to be visited only upon those whose lifestyles deserved it. Then, when some doctors failed to follow Ehrlich's instructions for the drug's use, with predictably disastrous consequences, he was accused of being a fraud. In the face of such accusations Ehrlich's health failed, and he died a broken man.

THALIDOMIDE Despite Ehrlich's efforts, the link between poison and medicine, between effect and side effect, remains. We can see it for ourselves on the list of possible side effects on every bottle of tablets, but it only really hits the headlines when something goes badly wrong. One of the worst such failures occurred in the latter half of the 1950s and early '60s, and the impact was so great that the name of the drug has gone down in history as synonymous with medical disasters: thalidomide.

The decision to ignore warnings about the safety of thalidomide is one reason that the drug appears in another book in this series: *History's Worst Decisions*. However, when it comes to inventions, it is certainly up there with the worst of them, and so is included here too.

First synthesised by chemists working for the German pharmaceutical company Grünenthal in 1954, thalidomide did not have the properties they were seeking but it was found to be an effective sedative. In fact, it seemed a miracle drug. Unlike the other sedatives in use at the time, it seemed to be completely safe. The drug was rushed onto the market in Germany in 1956 under the name Contergan; it was released a year later in the United Kingdom as Distaval.

Between 1956 and 1961, the drug was marketed in 50 countries. In the next few years, thalidomide was combined with other drugs and prescribed for a wide variety of conditions. One major group of users was pregnant women, who took thalidomide to combat morning sickness and as a 'safe' sedative. Unfortunately, adequate trials had not been carried out, and research into its effects on the development of embryos was only completed after the drug had been licensed for human use.

The first case of a 'thalidomide baby' occurred in 1956, when a child was born with severe malformations of the ears. From 1960 German doctors began to notice an unusually high number of cases of children born with defects of the limbs and ears, and other developmental problems. The numbers were so large that one doctor talked of an 'epidemic'. The link with thalidomide was established independently in both Germany and Australia in 1961. Soon supporting evidence was coming in from countries in Europe, Asia, North and South America and the Middle East, where thalidomide had been released for human use.

There were only 17 cases in the United States, thanks to the efforts of the Food and Drug Administration's Dr Frances Kelsey (b. 1914), who ensured that approval for the drug was withheld. The greatest number of cases occurred in the former West Germany and in the United Kingdom. In Germany thalidomide was available over the counter without a prescription; and in Britain, where it was a prescription-only drug, it was widely prescribed to mothers for morning sickness during early pregnancy. From 1956 to 1962, an estimated 10,000 to 12,000 children were born with severe birth defects because their mothers had taken thalidomide during the first few weeks of pregnancy. Approximately 40 per cent of these children died before their first birthdays.

DR FRANCES KELSEY
Dr Frances Kelsey, the Food and Drug Administration medical officer who kept thalidomide off the American market, appearing before a Senate sub-committee on August 1, 1962.

Thalidomide is generally associated with the missing and damaged limbs of many thalidomide survivors. However, the drug can affect almost all the organs of the body. A second major series of malformations affects the ears, eyes and eye muscles, facial nerves and muscles, and tear ducts. Victims can also suffer from serious internal problems affecting the heart, urinary tract, alimentary canal and reproductive system.

Research established that there was a specific period of sensitivity during which different developmental effects took place. Thalidomide only produced malformations in the foetus if it had been taken after the 34th day after the mother-to-be's last menstruation and before the 50th day. Within the two-week sensitive period from the 35th to the 49th days, different problems occurred depending on exactly when the drug had been ingested. If the mother took thalidomide between the 35th and 37th days, the effect was an absence of the ears and profound deafness; between the 39th and 41st days, the arms were missing; between the 43rd and 44th days, phocomelia (absence of the long bones in the arms but with digits attached to the shoulder joint) with three fingers; and between the 46th and 49th day, a three-jointed thumb. If thalidomide has been taken throughout the sensitive period, the damage to the limbs and internal organs was so severe that it often led to the early death of the child. In addition, thalidomide victims exhibited a range of neurological and developmental problems, including learning difficulties, dyslexia, autism and epilepsy.

In every country where thalidomide had been released for human use, the epidemic of limb and ear problems followed the sales of the drug with an eight-month delay. Thalidomide was withdrawn in Germany at the end of 1961, and the epidemic ended as predicted in the summer of 1962. In Japan, where thalidomide was withdrawn in fall 1962, the incidence of birth defects ended later than in Germany but also eight months after withdrawal. The same pattern was recorded in Ireland, Italy, the Netherlands, Sweden and the United Kingdom.

THALIDOMIDE ON TRIAL The public prosecutor's office in the German city of Aachen began work on a dossier on thalidomide at the end of 1961. By 1968 it had prepared an indictment of 972 pages, based on more than 500,000 pieces of evidence. The trial of seven Grünenthal executives began in May 1968. They were accused of selling a drug that caused severe physical harm without adequate testing, of failing to respond to data about side effects quickly enough, and having tried to suppress the information. The trial continued for two years and seven months but was halted when Grünenthal reached an out-of-court settlement with the German thalidomide victims. British, Canadian, Japanese and Swedish victims received similar compensation after their own legal proceedings. Sadly, the after-effects continue to the present day.

The thalidomide tragedy was a wake-up call to regulatory authorities the world over. It led to new testing protocols for drugs to ensure that they were safe for women and their unborn children. It taught people that the brave new pharmacological world that promised the dream of a future without pain or disease was just that – a dream.

HIGH STEPPERS: PLATFORMS AND HEELS

FAILING

Never got off the drawing board

Didn't work in practice

Killed its inventor

A commercial failure

Unforeseen consequences

Was used for evil ends

A success born of failure

Main Culprits: Footwear industry and celebrities

Motivation: Fashion

Damage Done: Ridicule; injuries to legs, feet and spine; accidents

'My platform shoes from Saturday Night Fever are in my closet. I don't wear them, but once in a while I peek at them.'

John Travolta, *Premiere* magazine

© 350jb | Dreamstime.com

It seems that the extremities of the human body are particularly prone to strange fashion crazes. We examined eighteenth-century dress wigs earlier (see pp. 85–9), and now it's the turn of footwear.

Perhaps the worst example is the traditional Chinese custom of foot-binding, which continued well into the twentieth century. The practice of deforming the feet of women to make them even more dependent on their menfolk strikes the modern reader as barbaric. Who would think of confining their feet in footwear that is so cramped that they have to totter around precariously, risking life and limb at every step, just to make themselves more attractive to the opposite sex?

WALKING TALL The elevated shoe is not a recent idea; the ancient Greeks and Romans favoured thick-soled shoes for theatrical performances, though in daily life they preferred to go barefoot or in sandals and buskins. In sixteenth-century Venice, courtesans wore an elevated outer shoe called the *chopine*, which not only gave them added height, but also protected their more delicate indoor footwear from the dirt of the unpaved streets. Japan's traditional *geta* clogs, still worn by the geisha and maiko of twenty-first century Kyoto, serve the same purpose.

The platform shoe has had several periods of popularity in the Western world. The first twentieth-century designs appeared in the 1930s and '40s, but the height of the fad was in the 1970s, the era of funky British pop fashion and glam rock. While earlier versions of the platform shoe had been strictly for women only, young men took to them after actors and pop musicians of the day, including John Travolta, Elton John, David Bowie and the band Kiss, wore them for their performances. Soon platform shoes made the transition from the disco to every walk of life, from the school playground to the factory.

However, in addition to being ungainly and ugly, platforms were also dangerous. High platforms were a health hazard for wearers who were more likely to trip and fall because the thick soles insulated their feet from the ground, affecting their perception of balance. Doctors also expressed fears about spinal problems caused by prolonged wear.

The fad for platforms petered out in the United Kingdom in the late 1970s and lasted a few years longer in the United States, but the day of the platform shoe was far from over. Along with much of 1970s fashion, platforms made a comeback in the 1990s, when British fashion designer and eccentric Vivienne Westwood featured them in her collections. A few years later, they were adopted by the girl-band the Spice Girls, which sealed their popularity with young women but less so, this time, with young men. Several different styles followed: from the clumpy buffalo trainers and boots to the clear plastic platform shoes favoured by pole dancers and strippers, which are now increasingly popular with female celebrities.

The alternative elevated shoe is the high heel that raises the heel higher than the toes, and which along with extra height gives the illusion of a longer, slenderer and more defined calf. High heels were in vogue for both men and women in the royal courts of Europe during the eighteenth century – the time to which we owe the expression 'well-heeled', meaning smartly turned out and well to do. When men and women's dress was simplified after the French Revolution of 1789, high heels vanished for a century to make a reappearance in late Victorian times. In the past 50 years, heels have risen and fallen with the dictates of fashion: high in the 1940s and '50s, low in the 1960s and '70s, and soaring upward again since the 1980s. A trawl through the websites of the fashionable shoe retailers reveals that a 2½ in (6 cm) heel is considered a 'low heel', a 2½ to 3½ in (6–8.5 cm) is a 'mid heel', and anything higher is a 'high heel'.

Podiatrists reveal that 75 per cent of their business from women is due to high-heeled shoe-related problems. High heels cause a wide range of conditions, including foot pain and foot deformities such as bunions and hammertoes; they cause an unsteady gait, leading to falls and twisted ankles; they shorten the wearer's stride and make it difficult for the wearer to run or walk on uneven surfaces; the position of the foot in the shoe predisposes regular wearers to degenerative conditions of the knee joint; the downward angle of the foot causes the Achilles tendon to shorten, which can lead to problems when walking in flat shoes; and much greater weight is transferred to the ball of the foot, increasing the likelihood of damage to the soft tissue that supports the foot. Finally, high heels – because they tip the foot forward – put pressure on the lower back and push the rump outward, compressing the lower back vertebrae and shortening the muscles of the lower back.

PODIATRISTS REVEAL THAT 75 PER CENT OF THEIR BUSINESS FROM WOMEN IS DUE TO HIGH-HEELED SHOE-RELATED PROBLEMS.

CATCH ME IF YOU CAN: THE BIRTH OF THE COMPUTER VIRUS

Main Culprits: Anonymous computer geeks

Motivation: Malice and egotism

Damage Done: Billions of dollars in lost data and business and clean-up work

'Since 1999, with some degree of regularity, a series of worms/viruses has swept around the world, crashing and infecting both commercial and home systems. A system of warnings and corrective patches to download have been put in place and reporting sites have proliferated. It seemed to begin with the Melissa virus in March 1999, followed by BubbleBoy in November. May of 2000 witnessed the spread of the I Love You worm, which had twice the impact of Melissa.'

Richard Rosenberg, *The Social Impact of the Internet,* **2004**

When you and I catch a bug – a bacteria or virus – we either take a few days off or go to see a doctor, but we don't imagine that the little critter has a personal grudge against us and made a point of choosing us to infect. It's just another microorganism looking for a nice warm place to set up home and start a family. The 'computer virus', in contrast, is a small and often very ingenious piece of software that has been designed by another human person to infect a file, computer application or operating system.

Computer viruses range from the stupidly cute, randomly printing a dumb phrase or pasting an image on your screen, to the viciously destructive, corrupting data files, wiping hard drives, irreparably crashing operating systems and applications and even destroying hardware, generating an annual clean-up cost of billions of dollars. In the early days of computing, viruses were spread from computer to computer through removable media – floppy disks and the like – so they could only travel a certain distance before the infection was identified and isolated. But with the advent of computer networks and the Internet, viruses have gone global. By using email, they can be unwittingly spread across the planet in minutes.

Readers below the age of 30 may not remember the 5,000 or so years during which human civilisation rubbed along quite well without the vital contributions made by email, search engines and online porn, but such a fabled time did indeed exist. It was as recently as 1969 that the first computers crawled out of the primeval ooze, looked up at the stars and dragged themselves up the beach of connectivity to establish the first computer 'network', the ARPANET (Advanced Research Projects Agency Network). The network that operated from 1969 to 1983, when it was absorbed into the Internet, spawned some rather spectacular innovations that still feature in our current global networks: email in 1971, File Transfer Protocol (FTP), in 1973 and, of course, the computer virus.

EARLY WARNING

The first virus, fairly innocuous by today's standards, displayed the message 'I'm the creeper: catch me if you can' on the machines it infected. The culprit was never identified and it is likely that he or she wrote the 'Reaper' program that sought out and deleted the Creeper virus to atone for his or her original misdeed.

Many early viruses were written as clever jokes and did little more than display absurd messages on computer screens. The ANIMAL virus of 1975, for example, was a type of simple game that asked questions to guess what animal a user was thinking of. The program was specifically designed not to damage files or root directories, and did not replicate itself maliciously on other computers. Other viruses, however, were written to be intentionally malicious and destructive. The 'Rabbit' virus of 1974 clogged a computer's operating system by replicating itself at speed, much like a biological virus inside a human body.

The first major computer virus outbreak on Apple Macs was caused by Richard Skrenta's 1982 Elk Cloner virus, which infected files in the Mac's operating system. Frederick Cohen came up with the term 'computer virus' in 1984, defining it as 'a program that can "infect" other programs by modifying them to include a possibly evolved copy of itself'.

IN THE WILD A virus 'in the wild' is one that has spread beyond its original computer or network to infect other machines. The Farooq Alvi brothers of Lahore, Pakistan, claimed the title of the first IBM-PC compatible virus in the wild with their 'Brain' virus of 1986. This was followed by other fancifully named creations, which finally spawned the creation of the first anti-virus software by IBM Corp in 1988. From then on, a worldwide electronic war has been waged between the virus creators and their foes, the anti-virus software developers.

Before the Internet, viruses were spread through removable media, the most common of which was the floppy disk. Viruses were attached to applications or data files on the disks, or to the boot sector of the disk, so that the virus would be uploaded when the computer 'booted up'. With the demise of the floppy and the appearance of more secure forms of removable media, such as the read-only CD, viruses had to find new ways of infecting PCs.

Viruses, although certainly dangerous, do not always live up to the considerable hype that they generate. Such was the case of the Michelangelo MS-DOS boot sector virus, which according to media reports was supposed to trigger a worldwide electronic apocalypse on March 6, 1992, by wiping the hard drives of every MS-DOS computer on the planet. However, when the fateful date arrived the problems caused were fairly minor in scale, and fewer than 20,000 cases of lost data were reported.

In the mid-1990s a new family of 'macro' viruses began to infect Microsoft Office programs such as Word and Excel. The first of these was the 'Concept' virus, created in 1995, which infected Word documents. With the huge growth of the Internet in the late 1990s, it soon became a major conduit for disseminating viruses and other horrors, such as 'worms' and 'Trojan horses'.

A worm differs from a virus in that it does not need an existing program to replicate itself. Instead it uses a network to send copies of itself to other terminals. A worm can take up so much bandwidth that it will significantly slow down or crash a network. David Smith's 'Melissa' worm caused havoc when it was released in 1999. Named for a Florida lapdancer Smith had taken a shine to, the Melissa worm was spread through Word and Excel and could email itself through Microsoft Outlook. Smith, however, did not escape retribution. He was tracked down by the FBI; was tried, convicted and sentenced to 20 years in jail.

DAVID SMITH
David L. Smith photographed leaving the Federal Court in Newark, New Jersey, after pleading guilty to knowingly spreading the infamous 'Melissa' computer virus in 1999 with the intent to cause damage.

One of the most infamous worldwide computer infections of recent times was the 'I Love You' worm of 2000, which originated in the Philippines. The virus began to spread on May 4. It arrived in an email with the suggestive subject line of 'I LOVE YOU' along with an attachment. When the attachment was opened on an MS Windows machine, the bug sent a copy of itself to everyone on the user's address list, and it also made several damaging changes to the user's operating system. Technically the virus was fairly simple, but nevertheless it proved to be extremely effective. As the emails appeared to come from known contacts, and the subject line was intriguing to say the least, recipients were more likely to open the attachment.

Within a single day, I Love You had infected an estimated 10 per cent of the computers linked to the Internet worldwide. Government agencies and major corporations had to shut down their networks to eliminate the virus, and the estimated cost was in the range of US $5.5 to

US $10 billion. However, in spite of the chaos and the costs, because the Philippines had no anti-virus writing legislation at the time, no one was ever convicted for releasing the virus on an unsuspecting world.

The newest sub-family of computer viruses are the 'cross-site scripting' (XSS) viruses that use WWW applications to disseminate themselves. They add malicious scripts written in HTML and another scripting language into web pages in order to obtain access privileges to sensitive data, session cookies, and a variety of other website objects. Prominent names targeted include the search engine Google, email servers Google Mail and Yahoo!, the pay site PayPal and the reference site Wikipedia.

As computer networks grow ever larger and more complex, carrying vast amounts of sensitive data about each and every one of us, not to mention most of the world's cash, the threat from computer viruses grows ever greater. But for the time being, the net's Wyatt Earps are still getting the better of its Billy the Kids.

FORMAT WARS:
THE FAILURE OF BETAMAX

Main Culprit: Sony Corp

Motivation: Business innovation

Damage Done: Billions of wasted dollars

'The format war between Betamax and VHS became a "winner takes all" situation, and its aftermath haunted the electronics landscape in the following decades.'

Frederick Wasser, *Veni, Vidi, Video,* 2001

As the quote above makes clear, format 'wars' are as old as technology itself. For centuries, a large part of technological innovation was actually driven by warfare. We've seen the advances made in metal casting, chemistry and ballistics that were triggered by the development of gunpowder and gunpowder weapons (see pp. 38–43), and more recently those born of the development of nuclear physics (see pp. 203–08). Alongside military hardware, the development and sale of industrial consumer goods, especially since the beginning of the twentieth century, has been a proving ground for rival inventions. Sound recording has been a particularly contentious field, from the cylinder versus disc at the beginning of the century to the eight-track versus cassette in the 1960s and the various rival digital formats of today. In the mid-1970s, the now iconic scrap over video formats pitted two giants of the Japanese electronics industry against each other in a struggle for world market domination.

© Bratan | Dreamstime.com

VIDEO FORMATS
Video cassettes
(VHS, Betacam, S-VHS
and MiniDV).

In the late 1970s and '80s, every home had one or more VCR (video cassette recorder). They were the vanguard of the video revolution that has come to fruition today with cellphone cams and video websites such as YouTube. Prior to the commercialisation of the first home VCRs in 1972, 'home movies' meant exactly that – cine film that turned Mum and Dad into Cecil B. DeMille. Cine film was fine for recording cousin Jo's wedding or the family's summer holiday, but what consumers really wanted was something with which they could record their favourite TV shows and feature films.

For a generation weaned on television, the market for an affordable home-recording system was simply astronomical, and marketing executives worldwide were having heart palpitations just thinking about it. The Dutch electronics giant Philips had achieved a phenomenal commercial success with compact audiocassettes in 1963, and it was hoping to repeat its success with video. Philips developed the first home VCR system, which it launched as the 'N1500' on the UK market in 1972. However, the true muscle in the electronics industry in the 1970s had shifted eastward to the commercial and technical headquarters of Tokyo and Osaka. Two Japanese firms, Sony Electronics and the

Victor Company of Japan, better known as JVC, were squaring off for a battle worthy of any Japanese *King Kong versus Godzilla* B-movie.

In 1974, both Sony and JVC were ready with their prototype VCR formats. At a meeting of manufacturers that year, Sony demonstrated its 'Beta' format, pushing to get it adopted as the industry standard with the backing of the Japanese Ministry of Trade and Industry. At a later meeting, JVC and its parent, Matsushita Corp, unveiled the VHS ('Video Home System') format for the first time, but at that point Sony was ready to go into production with Betamax, making a format war unavoidable. Betamax launched in 1975, and VHS followed in 1976 in Europe and Asia and 1977 in the United States. By releasing its format first, Sony hoped to establish it as the industry standard. It is often claimed, especially by those who had invested in a Betamax VCR, that the format was far superior to VHS. If it really was, why did it fail?

The first major difference was recording time. The original Sony Betamax VCR could only record for one hour. The first JVC doubled that to two hours and with the collaboration of Radio Corporation of America (RCA) soon increased it to four. RCA had first been interested in Sony's Betamax but had concluded that the recording time was much too short. They wanted a format that could record an average four-hour American football game. The increase in recording time, however, had a cost in that the picture and sound quality were compromised. Sony bet that consumers would prefer shorter but higher-quality sound and pictures, while RCA, JVC and Matsushita were happy to trade quality for the extra recording time.

The quality of the initial one-hour Betamax format released in 1975 was better than VHS. It had a higher screen resolution, less colour variation and better sound quality than the VHS playback mode. However, as the war heated up, Sony had to sacrifice these advantages to keep up with the advances made by VHS. When it introduced Betamax's two-hour mode to compete with VHS's standard playback time, it reduced the Betamax screen resolution to match that of VHS. In the area of sound quality, Sony also maintained a lead with its Beta HiFi system, but here, too, JVC responded with its own VHS HiFi standard. Again, the quality differential between the two formats was negligible for the vast majority of end users.

HOME MOVIES

IN 1975, SONY HAD 100
PER CENT OF THE VCR
MARKET WORLDWIDE; BY
1981, THIS HAD SHRUNK
TO 25 PER CENT IN THE
UNITED STATES AND 7.5
PER CENT IN THE UNITED
KINGDOM.

In 1975, Sony had 100 per cent of the VCR market worldwide; by 1981, this had shrunk to 25 per cent in the United States and 7.5 per cent in the United Kingdom. Even when Sony released Extended Definition Betamax in 1988, with a screen resolution double the existing Betamax and VHS standards, it was too late; the war had been lost. That same year, Sony threw in the towel and began producing VHS VCRs of their own under licence. Betamax continued to exist in high-end formats, but the last Sony Betamax VCR was made in 2002. The last JVC VHS VCR rolled off the production line in 2007.

Sony had admitted defeat in the VCR format war, but it was determined not to be beaten again in the latest war over the successor format for DVDs – a contest between Toshiba's HD DVD and Sony's own Blu-ray. This latest format war ranged the biggest names in electronics, entertainment and software in a three-year contest from 2005 to 2008. Again, Sony supported a higher-quality and more expensive technology, but on this occasion, it was victorious. Toshiba announced that it was discontinuing the HD DVD format in February 2008.

I DID IT MY WAY: KARAOKE

FAILING

Never got off the drawing board

Didn't work in practice

Killed its inventor

A commercial failure

Unforeseen consequences

Was used for evil ends

A success born of failure

Main Culprit: Daisuke Inoue (b. 1940)

Motivation: Entertainment

Damage Done: Crimes against music

'Japanese technologies have empowered the "little guy": suddenly anyone can listen to his or her music while on a crowded train, fax their handwriting across the globe or perform their own rendition of "I Saw Mommy Kissing Santa Claus". Karaoke makes no one marginal.'

Xun Zhou and Francesca Tarocco, *Karaoke*, 2007

The Japanese have given the world many cultural treasures: the traditional arts of ink-wash painting, haiku poetry and calligraphy; the tea ceremony; minimalist architecture and interior design; and stone gardens. What these things share is a plainness bordering on austerity, which for many Westerners represents the epitome of restrained good taste. So we must ask why, after centuries of refinement, toned-down elegance and sobriety, did the Japanese unleash upon an unsuspecting world the musical horror that is karaoke?

Japanese cities are studded with tiny bars, most of which are barely larger than their counters and can accommodate no more than a dozen customers. It is in this rather cosy, intimate and unthreatening atmosphere – quite unlike the large, teeming, noisy and often rowdy bars of the United States and Europe – that karaoke (from the Japanese *kara*[*ppo*], 'empty', and *oke*[*sutora*], 'orchestra') first saw the light of day. A jobbing musician by the name of Daisuke Inoue (b. 1940) often accompanied the patrons of a bar in Kobe, a large, well-to-do port city in western Japan not far from Osaka, in their renditions of popular Japanese and Western ballads. His forte, so it is said, was to make even the most vocally challenged singer sound passable, by following rather than leading and turning up the volume to drown out the worst of the caterwauling.

© Michael S. Yamashita | Corbis

TWIN PASSIONS
Japanese office workers combine their twin national passions of karaoke and cherry blossoms at a cherry-blossom viewing party in Fukuoka, Japan, in 1993.

His services were much in demand, but he was not always available. In 1971 the director of a local firm asked him to record a backing tape for a company trip to an *onsen* (hot spring) resort. Instead of selling him a soundtrack, Inoue invented a playback machine consisting of a car stereo, an amplifier and a switch operated by a ¥100 coin (roughly US $1). The machine was a hit, and soon Inoue was making more backing tapes and machines to lease to bars in Kobe. The idea soon spread to other Japanese cities, and became a staple of the salaryman's (the Japanese term for an office worker) night out. Once nicely tanked up on *mizuwari* (a weak mixture of whiskey and water) or *biru* (lager), the patrons of the bar would serenade one another with renditions of their favourite songs.

Unfortunately, Inoue never patented his invention. If he had, he'd be one of the richest men in Asia by now. He remained largely unknown in the West until 1999, when *Time* magazine called him one of the twentieth century's most influential Asians, eulogising rather improbably that he 'had helped to liberate legions of the once unvoiced – as much as Mao Zedong or Mahatma Gandhi changed Asian days, Inoue transformed its nights'. In 2004, Inoue was awarded Harvard University's joke award the 'Ig Nobel' Peace Prize for inventing karaoke. The panel, which included real Nobel Prize winners, declared that he had invented 'an entirely new way for people to learn to tolerate each other'. At the presentation ceremony, Inoue received a standing ovation and regaled the audience with 'I'd Like to Teach the World to Sing'.

As Inoue had never taken out a patent, the Filipino inventor Roberto del Rosario took one out on his own 'Minus-One' karaoke machine. Following a court battle with a Japanese company that claimed to have invented karaoke machines first, del Rosario won his patents in 1983 and 1986.

PATENT WARS

In the 1980s, karaoke spread from Japan to its near Asian neighbours: Korea, Taiwan, Hong Kong, China and Singapore; by the 1990s, it had reached the rest of the world. Each region, however, has a distinct style of karaoke. In its country of origin, karaoke originally was perpetrated only in bars. However, this meant that it was only available to adults, usually male, late at night, and at expensive rates – as Japanese bars are not cheap. In order for karaoke to become truly universal, it had to become a form of entertainment for all the family. This was achieved not in Japan but in Hong Kong in the 1980s (then British, now Chinese), where the first 'karaoke boxes' were devised.

A karaoke box is a cramped, soundproofed room, equipped with seating and a karaoke machine. Imagine the cell for the electric chair and you won't be far off. The victims, I mean, patrons, rent the room by the hour. Karaoke boxes are now the most popular format in Asia, which means that one can now find bars in Hong Kong, Tokyo and Shanghai to enjoy a drink without being serenaded by inept, heavily accented versions of 'My Way' and 'Bohemian Rhapsody'. Karaoke is extremely popular in South Korea, but was banned in North Korea in 2007. The world's last remaining Marxist-Leninist dictatorship explained that its action was intended to 'crush enemy scheming and to squarely confront those who threaten the maintenance of the socialist system'.

In the West, in contrast, karaoke is a much more public phenomenon. Although home karaoke machines have been available since the 1990s, they never really took off in the North American and European markets. Only recently have karaoke-style games for console systems made the youth of the United States and Europe put down their hairbrushes and pick up their microphones.

Outside the home the most popular options are dedicated karaoke venues and karaoke nights in bars, clubs, pubs and restaurants. In dedicated venues, participants perform on stage often with professional mics, disco lighting, dance floors and multiple screens displaying the lyrics. There is usually no charge or a token fee per song, as the establishment makes its money by selling food and drinks.

Karaoke is now ubiquitous: in addition to karaoke boxes and bars, it is available through home-movie systems, on cellphones and through the Internet. This explosion of singing opportunities, however, does not seem to have noticeably improved the musical abilities of the general population. Quite to the contrary, if shows such as *American Idol* and *X-Factor* are anything to go by, karaoke, by massaging and improving on people's feeble efforts, has given many the delusion that they can sing and perform. But readers should remember that, even with considerable natural talent, singing is a skill that needs to be nurtured and developed over years. Doing it 'your way' isn't going to turn you into the next Frank Sinatra but just make you the butt of some cruel judge's jokes.

ELECTRONIC GARBAGE: THE DISSEMINATION OF SPAM EMAIL

FAILING

Never got off the drawing board

Didn't work in practice

Killed its inventor

A commercial failure

Unforeseen consequences

Was used for evil ends

A success born of failure

Main Culprit: Gary Thuerk of Digital Equipment Corp

Motivation: Greed

Damage Done: Billions in lost time, clogged email servers and financial scams

'… spam is worse than irritating. It is a drain on business productivity … spreads scams, pornography, and even computer viruses … [and] preys on less sophisticated email users, including children, threatening their safety and privacy … In short, spam threatens to undo much of the good that email has achieved.'

Bill Gates, *The Wall Street Journal*, June 23, 2003

© Instamatic | Dreamstime.com

The Internet has transformed our lives for the better in a multitude of ways: by giving us access to services and information in our homes and offices that our parents' generation could only dream of, and a new means of keeping in touch with family, friends and colleagues far and near. Unfortunately, it has also made us accessible to some much less desirable electronic correspondents – from the annoying but innocuous advertiser to the sophisticated criminal scammer who is after our money or identity. We've already examined the willfully destructive 'computeristas' who create computer viruses (see pp. 226–30), and now it is the turn of that other Internet bugbear: electronic junk mail, or 'spam'.

The first spam appeared on Usenet newsgroups but quickly graduated to email. It can now be found across the full range of electronic media: instant messaging, chat rooms, forums, mobile phones, online gaming, search engines (spamdexing), reference sites (wikispam), blogs (blam), social networking and video-sharing sites.

SPAM, SPAM, SPAM AND SPAM The first bulk unsolicited email appropriately involved a computer marketeer and predated the invention of the Internet as we know it today. In 1978 Gary Thuerk, a marketing manager with Digital Equipment Corp, hit upon the idea of informing the 6,000 West Coast subscribers to the US Department of Defense's ARPANET about new computer equipment. Instead of emailing members of the network one by one, he discovered a way of sending them all the same email – bulk emailing had been born. The recipient reaction to this first ever case of spam was predictably hostile.

Although there were a few spam-like outbreaks during the 1980s on Usenet, the first major commercial spam incident occurred in 1994, when immigration lawyers Laurence Carter and Martha Siegel advertised their services via bulk Usenet postings in a case that became known as the 'green-card spam'. Within a few years, spam had migrated to its current preferred medium, email.

SPAM™, as many readers will know, is a brand of processed and canned meat made by the Hormel Food Corporation of Austin, Minnesota. It famously featured in a *Monty Python's Flying Circus* sketch of 1970, in which customers in a diner are offered dishes that all contain SPAM™: 'Well there's … egg, bacon and spam; egg, bacon, sausage and spam;

spam, bacon, sausage and spam; spam, egg, spam, spam, bacon and spam; spam, spam, spam, egg and spam; spam, spam, spam, spam, spam, spam, baked beans, spam, spam, spam and spam ...' and so on. The Pythons poked fun at SPAM™'s blandness and wartime popularity, when meat was in short supply but SPAM™ itself was never rationed.

Spamming began in the 1980s, targeting newsgroups and chat rooms, and the term was first applied to nuisance postings in the early 1990s. Because connection speeds were extremely slow in the early days of the net, spam messages took many minutes to download, gumming up the electronic works. The term finally made it into the *Oxford English Dictionary* in its 1998 edition. Hormel Foods fought a rearguard action to try and prevent the association of their trademark with one of the most unpopular practices of the Internet age. However, repeated judgments have confirmed that spam is now considered a generic term that cannot be trademarked. Hormel laconically stated on its website: 'We are trying to avoid the day when the consuming public asks, "Why would Hormel Foods name its product after junk email?"'

Spam began life as a nerdy practice that affected just a few thousand Usenet aficionados, but with the appearance of the Internet in every home and office, the problem was transformed into a global problem of gargantuan proportions. As a daily computer user, I am probably not untypical in receiving somewhere between 20 and 30 pieces of unsolicited email in my mailbox every day. Some of it is addressed to me directly; some of is addressed to people who have a similar name or share my Internet Service Provider (ISP); and, more worryingly, some of it seems to come from myself. It is estimated that spam email now comprises between 80 and 95 per cent of all email traffic worldwide. I have noticed that the types of spam that my acquaintances and I receive seem to vary. A few years ago, there were a lot of links to porn sites, then there was a flurry of strangely worded stock-market tips and bogus academic degrees; at the time of writing the latest crop seems to split between ads for anti-impotence medication and designer watches – all fake, no doubt.

Although spam has been outlawed in many jurisdictions, it continues to be economically viable to advertisers because they can reach millions of customers at very little cost to themselves when compared to

IT IS ESTIMATED THAT SPAM EMAIL NOW COMPRISES BETWEEN 80 AND 95 PER CENT OF ALL EMAIL TRAFFIC WORLDWIDE.

conventional advertising. With the numbers involved, it requires only a small take-up of any product to make spamming worthwhile. Any cost is borne by the public and the ISPs. Increasingly, spam emails are sent via 'zombie networks' of 'slave' machines that have been infected with a piece of malware (a virus, worm or Trojan horse). Unbeknown to the owner, who usually leaves his or her computer on and linked to the Internet at all times, the spammer can access his or her email software and use it to send out literally millions of emails.

According to recent research, the United States leads the table for originating spam, with 28 per cent; it is followed by South Korea, 5 per cent; China, 5 per cent; and Russia and Brazil, with 4 per cent each. However, this geographic spread hides one of the most extraordinary aspects of spamming: how few people it takes to generate the billions of emails that deluge the planet every year.

In October 2008, the Atkinson brothers based in the country town of Christchurch, New Zealand, were the first spammers convicted under the provisions of the Unsolicited Electronic Messages Act (2007). Evidence submitted at their trial estimated that, at its height, the Atkinson brothers' operation was responsible for as much as one-third of the world's spam. The EU estimated the total cost of spam in 2001 to around US $13 billion dollars, and a 2007 report by the state of California quoted the same figure for costs incurred in the United States alone. The costs in wasted time and extra bandwidth, software and human resources is borne by governments, corporations and ISPs, but ultimately it is us as end users who pay for it in increased taxes and prices.

GONE PHISHING The term 'phishing', meaning the practice of acquiring sensitive information such as usernames, passwords and bank and credit card accounts through deception by email, entered the English language in 1996. The practice, however, has been around since 1987, when the first phishers targeted AOL users, pretending to be staff and asking their victims the old stand-bys, to 'verify their accounts' and 'confirm their billing information', in order to steal their credit card numbers. By 1995 AOL had dealt with the problem, and the phishers headed for pastures newer and greener.

Phishing now targets the full range of financial institutions from banks and online tax payment services. A common early tactic mimicked the AOL scam, by sending an email purporting to be from the victim's bank, and asking him or her to enter sensitive information. However, by now, most bank customers have been warned off this simple ploy, so the phishers have adopted more sophisticated techniques. Another scam involves creating a facsimile of a financial institution's website. Doing this is actually far simpler than it seems, as the source code of websites and their graphics are not difficult to download and copy. Then the scammers use links embedded in emails to redirect victims to the fake site.

Losses from phishing have been fairly spectacular, and at little risk to the criminals who, unlike bank robbers, do not have to brave armed security guards and law enforcement officers in hot pursuit. Estimates for losses incurred in the United States from phishing attacks in 2007 are in the order of US $3 billion plus.

FAILING

Never got off the drawing board

Didn't work in practice

Killed its inventor

A commercial failure

Unforeseen consequences

Was used for evil ends

A success born of failure

ELECTRIC DREAMS: HUMAN–ELECTRIC HYBRID VEHICLES

Main Culprit: Sir Clive Sinclair (b. 1940)

Motivation: Scientific innovation

Damage Done: Slowed the acceptance of electric vehicles in the United Kingdom for a decade

'No amount of persuasion would force the public to admit that they had a need for a car that looked more like a pedal car, and travelled little faster than a bicycle. Moreover, only one in ten new consumer goods survives in the marketplace. The public are not gullible fools!'

Fiona McLean, *Marketing the Museum*, 1997

Transport has always been a fertile ground for innovation. In the previous pages we have seen quite a few 'planes, trains and automobiles', or their approximations, at least. But this being a book about history's worst inventions, few of these designs have proved successful. In an era of worsening air pollution and escalating oil prices – during the 1970s and '80s – which also saw oil shocks and rising prices, the attraction of a clean, inexpensive vehicle that did not depend on petrol seemed irresistible. Although global warming and our carbon footprints were not yet major concerns, the solution hit upon by inventors of the 1980s tackled both head-on.

British inventor Sir Clive Sinclair (b. 1940) can claim several world firsts: the first slim-line pocket calculator and the first affordable mass-market home computer, as well as one conspicuous failure, the first commercial human–electric hybrid vehicle (HEHV), which combined an electric motor with pedal power. According to his autobiography, Sinclair had been dreaming of inventing an electric-powered vehicle since he was a teenager. In the early 1980s, having made his fortune in calculators, he finally got a chance to make his electric dreams come true.

CLIVE SINCLAIR
Clive Sinclair shows off his new pocket television with flat picture tube in London in 1981.

The new vehicle, the Sinclair C5, was a one-seater battery-assisted tricycle, which the driver steered by handles on either side of the seat, like the steering mechanism of some recumbent bikes. The C5 could operate on battery or pedal power alone, but its top speed was limited to 15 mph (24 km/h), this being the maximum speed allowed for a vehicle to be driven on British roads without a driving licence. Sinclair hired the Lotus sports and racing auto company to work on the production models, and the vacuum cleaner maker Hoover Ltd manufactured the vehicle. The C5 was launched in a blaze of publicity in January 1985, retailing at the affordable price of £399 (US $650).

The C5 was an instant and resounding flop. True, the C5 had some technical and design faults; it had neither a roof nor high sides to protect the driver from the British climate, and it was low to the ground, which raised fears that the C5 would be unsafe in heavy traffic. However, these problems have not put off cyclists, both conventional and recumbent,

THE LENGTH OF THE PEDALS' SHAFTS COULD NOT BE ADJUSTED TO THE DRIVER'S HEIGHT; THE COLD WEATHER SHORTENED BATTERY LIFE; AND THE ELECTRIC MOTOR OVERHEATED ON HILLS.

and motorcyclists who throng British roads in any weather. On the technical side, criticisms were more justified: the length of the pedals' shafts could not be adjusted to the driver's height; the cold weather shortened battery life; and the electric motor overheated on hills. But if the truth be told, the fundamental problem was the styling of the C5, which looked exactly like a child's plastic pedal car. Concerns over the environment were not as pressing as they are today, and this combined with its technical shortcomings meant that the C5 was doomed. Fewer than 20,000 models were sold, and in August 1985 Hoover halted production. That October, Sinclair Vehicles filed for bankruptcy.

Another vehicle saw the light of day in the mid-1980s; the TWIKE (short for 'two in a bike'). The pedal-powered TWIKE won its Swiss design team first place in the Human-Powered Vehicle Speed Championship at the 1986 World Expo in Vancouver, Canada. The vehicle was modified in the early 1990s with the addition of a battery-powered AC motor, and in 1995, the Swiss TWIKE Company rolled out the TWIKE III onto the roads of Europe.

The TWIKE is a more sophisticated vehicle than the Sinclair C5, and it does not suffer from its major faults. It still looks a bit like child's toy, but at least the driver and passenger are protected from the weather by a lightweight aluminium and plastic shell. The TWIKE weighs in at 542 lb (246 kg), too heavy to be driven by pedal power alone. The five-gear hub, however, does extend the range of the vehicle because it transfers its power directly into the drive train. In addition, a regenerative anti-lock braking system recharges the computer-controlled batteries during operation. The TWIKE has a respectable top speed of 53 mph (85 km/h) and a range of 100 miles (160 km). A full charge costs an average of 20–30 pence in the United Kingdom and 30–45 cents in the States.

The TWIKE has yet to achieve a runaway success. At the time of writing, fewer than 2,000 TWIKEs have been sold in Europe, where they retail at between 17,800 and 32,200€ (US $25,000–45,000), and in the United States, where the basic model costs US $35,000. However, with the rising financial and environmental costs of driving, and the world's major auto makers on the rocks, you may soon find yourself being overtaken not by a BMW, Toyota or Ford, but by a TWIKE.

NEITHER A BORROWER NOR A LENDER BE: SUBPRIME MORTGAGES

FAILING

Never got off the drawing board

Didn't work in practice

Killed its inventor

A commercial failure

Unforeseen consequences

Was used for evil ends

A success born of failure

Main Culprits: Lenders and borrowers

Motivation: Greed

Damage Done: Collapse of the world's financial system; global recession

'In fact, subprime lending has become so accepted that Fannie Mae and Freddie Mac, seeing that they've left a lot of business to subprime lenders, have gotten into the fray by introducing their own version of subprime loans. So if subprime loans have been "endorsed" by conventional lenders, then what's the matter with them? Actually nothing.'

David and Carl Reed, *Who Says You Can't Buy a Home!*, 2006

How wrong can you be? But that's a cheap jibe because how many among us – the Wall Street 'masters of the universe', the average man in the street or the chairmen of the Federal Reserve or the Bank of England – can truly claim to have seen it coming? If we're honest, most of us were far too busy working out how much money we'd made on house prices and in the stock market to heed the warnings. The primary cause of what will go down in history as the 'credit crunch' of 2008 was the same as in all preceding bubbles (see pp. 64–9): unrealistically inflated asset prices that suddenly lost their value. We saw it in the seventeenth-century Dutch 'Tulip mania', the eighteenth-century South Sea Bubble, the great crash of 1929 and the dotcom bubble of the turn of the millennium. However, many of the bubbles of the late twentieth and early twenty-first centuries have favoured house prices. As the new century got underway, banks, mortgage lenders, hedge funds and other financial institutions, as well as private investors and homeowners, made the frankly insane bet that house prices would continue on an upward trend indefinitely.

THE CRASH | The outcome was exactly the same as in previous bubbles – the sudden crash in asset prices leads to large-scale defaults by borrowers who have invested heavily in tulips, securities, dotcoms or houses, precipitating a banking crisis that itself triggers a recession in the wider economy. However, the most terrifying differences between the 2008 credit crunch and the recession it has engendered from the bursting of earlier bubbles are the speed at which the crisis developed and its truly global reach. In late 2006, there were warning signs of trouble in the US and UK housing markets, by 2008 the global banking system was on the verge of collapse, and all the world's major economies were slowing or going into recession.

With globalisation, no economy is safe. However, what makes the mess we have got ourselves into particularly difficult to sort out is the complexity of the interdependent financial instruments that have been traded worldwide for the past few decades of deregulation. Subprime mortgages are only the tip of a financial iceberg.

What has really done the damage is the repackaging and selling on of securities and derivatives based on these now 'toxic' mortgage assets, so that no one in the financial sector is actually sure of how much bad debt there is still to be uncovered. The credit-rating system has become null and void overnight. The result: no bank is willing to lend any more – to anyone.

In the Anglo-Saxon world, home ownership has always been a big deal, while in other developed economies people are happy to rent; hence, the housing market has always played a central role in the United States and United Kingdom's cycles of boom and bust. In a sense, today's credit crunch began back in 1938 during the previous major financial crisis to shake the planet, the Great Depression (1929–39). As part of the 'New Deal', President Franklin D. Roosevelt's 1933–45 administration created Fannie Mae (the Federal National Mortgage Association) to make home loans available to low-income American families. Then, in 1968, the federal government privatised Fannie and set up Freddie Mac (the Federal Home Loan Mortgage Corporation) in 1970 to provide competition.

F.D. ROOSEVELT
The American president whose 'New Deal' included the creation of Fannie Mae (the Federal National Mortgage Association) in 1938.

© Public Domain | Library of Congress Prints and Photographs Division

Back in the 1970s, a period that many people might think of now as a Golden Age – oil price hikes and an unpopular war and president surely sound familiar – Lewis Ranieri, a bond trader with the Wall Street investment bank Salomon Brothers, invented mortgage-backed securities (MBSs): five- and ten-year investment bonds that repackaged mortgages. Ranieri's invention, known as 'securitisation', began with home loans but has since been applied to all forms of debt, including personal loans and credit cards. At the time, he was hailed as one of the 'masters of the universe'. Today, he is just another casualty of the credit crunch, as his Franklin Bank went bust in 2008 at an estimated cost to the taxpayer of US $1.6 billion.

Ranieri may have invented MBSs, one of the several engines that have driven the subprime crisis, but he is by no means to blame for it. Until 1999 the US mortgage market was fairly well regulated and lending remained responsible. But President Clinton and then President Bush gradually took off the regulatory brakes, resulting in an increase in high-risk subprime lending. The first two years of the new millennium saw the bursting of the dotcom bubble and 9/11. Recession loomed, and the Fed and other central banks stimulated the economy by slashing interest rates. The world was awash with cheap cash ('liquidity' to an economist), which encouraged lenders, borrowers and investors to take much greater risks. In 2001 subprime mortgages totalled US $173 billion, but by the height of the house-price bubble in 2005, they reached a massive US $665 billion. However, because the economy seemed in good shape and house prices were rising, no one was worried.

Homebuyers gambled on ever-increasing house prices and mortgaged themselves to the hilt with non-traditional (subprime) mortgages, such as 2/28 interest-only loans, with a low introductory rate and no down payment. The US property bubble burst in August 2006. With low introductory interest rates coming to an end and falling house prices, homeowners could not renegotiate at a lower or equivalent rate, and they defaulted in their hundreds of thousands. So are homebuyers the people we should blame? In part, certainly, but the lending institutions also played their role with aggressive loan policies that persuaded their clients to borrow far more than they could afford.

© Delray | Dreamstime.com

WALL STREET
Located in lower Manhattan, New York, Wall Street is the heart of America's financial sector and the home of the 'masters of the universe', whose downfall has had dramatic consequences around the globe.

Had it stopped with borrowers and lenders, we might have experienced the bursting of yet another localised property bubble in 2006, with tragic outcomes for individuals who lost their homes and several overexposed banks and Savings-and-Loan associations, which went bust, and then probably a short recession. Unfortunately, through securitisation, the now toxic subprime mortgage assets have infected the whole of the global financial system. Investment banks have packaged mortgages as MBSs and collateralised debt obligations (CDOs), and credit rating agencies, which were charged with flagging up the soundness of financial products, gave them 'triple-A' ratings. As a result, individuals, banks and hedge funds worldwide poured billions of dollars into MBSs and CDOs.

The banks, which provide the credit for the world economy, suddenly found themselves holding trillions of dollars worth of assets that could not be accurately valued, causing a massive crash in confidence. The banks stopped lending to one another and to businesses, even those without any connection to the US subprime debacle.

At the time of writing, the world economic system has stalled, and no amount of pump priming by the US and EU governments has yet to make any difference. And according to Lewis Ranieri, this is only the beginning: 'If you think this is bad, imagine what it's going to be like in the middle of the crisis.'

THE LONG TERM

Taking the long view, this is the latest in a long line of recessions triggered by the bursting of an asset bubble. The world economy will recover, and a few generations will be a little bit wiser when it comes to borrowing and lending money on real estate. And then, taking the long view again, everyone will forget, and at some point in the future, there will be a new bubble, whether it's asteroid-mining stocks, robot company startups or real-estate deals on Mars that are too good to miss.

FURTHER READING

There are many good general history sources available, both in hard copy and online. For example, the *Encyclopedia Britannica*, which can be found both online at www.britannica.com and in print, is one of the best sources, while Microsoft's online encyclopedia, *MSN Encarta* can be found at encarta.msn.com

Please note that the listings below for specific chapters are intended to give you the background to the subject of the entry, and in several instances to be entry-specific references; some also go beyond what is included in the entries.

This resource includes web addresses that may change due to the constantly evolving environment of the Internet. If you find a website address that does not work, please try a keyword search, because the information will often still be available online, but will have moved to a different page.

Flights of Fancy: Icarus and the Dream of Human-Powered flight

Abbott, Allan and Wilson, David (1995) *Human-powered Vehicles*, Champaign, IL: Human Kinetics Publishers

Grosser, Morton (2004) *Gossamer Odyssey: The Triumph of Human-Powered Flight*, Osceola, WI: Zenith Press

Sherwin, Keith (1976) *To Fly Like a Bird: The Story of Man-Powered Aircraft*, Folkestone: Bailey Bros. & Swinfen

The Miracle Mineral: Charlemagne's Asbestos Tablecloth

Bowker, Michael (2003) *Fatal Deception: The Terrifying True Story of How Asbestos is Killing America*, New York: Rodale

Guthrie, George B. and Mossman, Brooke T., ed (1993) *Health Effects of Mineral Dusts*, Virginia: Mineralogical Society of America

Maines, Rachel (2005) *Asbestos and Fire: Technological Tradeoffs and the Body at Risk*, Chapel Hill, NC: Rutgers University Press

Black Gold: The Uses and Abuses of Petroleum

Hyne, Norman J. (2001) *Nontechnical Guide to Petroleum Geology, Exploration, Drilling and Production*, Tulsa, Pennwell Books

Heinberg, Richard (2005) *The Party's Over*, Forest Row, East Sussex: Clairview Books

Roberts, Paul (2005) *The End of Oil: On the Edge of a Perilous New World*, Boston, Mariner Books

Deadly Matters: The Origins and Development of Capital Punishment

Banner, Stuart (2003) *The Death Penalty: An American History*, Cambridge, MA: Harvard University Press

Costanzo, Mark (1997) *Just Revenge: Costs and Consequences of the Death Penalty*, USA: Worth Publishers

Zimring, Franklin E. (2004) *The Contradictions of American Capital Punishment*, New York: OUP USA

Going Up in Smoke: Native Americans Invent Smoking

Apperson, George (2006) *The Social History of Smoking*, Teddington: The Echo Library

Kluger, Richard (1997) *Ashes to Ashes: America's Hundred-Year Cigarette War, the Public Health, and the Unabashed Triumph of Philip Morris*, New York: Knopf

World Health Organization, Tobacco Free Initiative, http://www.who.int/tobacco/en/

Bang! Bang! You're Dead: The Invention of Gunpowder and Guns

Croll, Mike (1998) *The History of Landmines*, Barnsley: Leo Cooper

Kelly, Jack (2005) *Gunpowder: Alchemy, Bombards, and Pyrotechnics: The History of the Explosive That Changed the World*, New York: Basic Books

Ponting, Clive (2005) *Gunpowder*, London: Chatto & Windus

Inheritance Powder: Arsenic, the Poisoner's Favourite

Emsley, John (2006) *Elements of Murder: A History of Poison*, Oxford: Oxford University Press

Meharg, Andrew (2004) *Venomous Earth: How Arsenic Caused the World's Worst Mass Poisoning*, New York: Palgrave Macmillan

Tiptoe Through the Tulips: The Scourge of Landmines

Monin, Lydia and Gallimore, Andrew (2002) *The Devil's Gardens: A History of Landmines*, London: Random House UK

Ponting, Clive (2005) *Gunpowder*, London: Chatto & Windus

Williams, Jody (2008) *Banning Landmines: Disarmament, Citizen Diplomacy, and Human Society*, Lanham: Rowman & Littlefield Publishers

All Choked Up: Leonardo da Vinci Reinvents Chemical Warfare

Coleman, Kim (2005) *A History of Chemical Warfare*, Basingstoke: Palgrave Macmillan

Harris, Robert and Paxman, Jeremy (2002) *A Higher Form of Killing: The Secret History of Chemical and Biological Warfare*, London: Random House

Mauroni, Albert J. (2006) *Chemical and Biological Warfare: A Reference Handbook*, Santa Barbara, CA: ABC-CLIO

Rocket Man: China's First Astronaut

Sutton, George P. (2005) *History of Liquid Propellant Rocket Engines*, Reston VA: American Institute of Aeronautics and Astronautics

It's a Mad, Mad World: Tulip Mania and Other Financial Follies

Dash, Mike (2001) *Tulipomania: The Story of the World's Most Coveted Flower & the Extraordinary Passions It Aroused*, New York: Three Rivers Press

Goldgar, Anne (2007) *Tulipmania: Money, Honor, and Knowledge in the Dutch Golden Age*, Chicago: University of Chicago Press

Kindleberger, Charles P., Aliber, Robert and Solow, Robert (2005) *Manias, Panics, and Crashes: A History of Financial Crises*, Hoboken: Wiley

Cool, Fizzing, Tasty, Sweet: Comagnie des Limonadiers and Fizzy Drinks

Critser, Greg (2003) *Fat Land: How Americans Became the Fattest People in the World*, London: Penguin

Finkelstein, Eric A. and Zuckerman, Laurie (2008) *The Fattening of America: How The Economy Makes Us Fat, If It Matters, And What To Do About It*, Hoboken: Wiley

Cutting Edge: The Switchblade and Gang Culture

Hagedorn, John M. and Davis, Mike (2008) *A World of Gangs: Armed Young Men and Gangsta Culture*, Minnesota: University of Minnesota Press

Ritchie, R. and Stewart, R. (1997) *The Standard Knife Collector's Guide*, Paducah, KY: Collector Books

Poisoned Chalice: Biological Warfare

Harris, Robert and Paxman, Jeremy (2002) *A Higher Form of Killing: The Secret History of Chemical and Biological Warfare*, London: Random House

Mauroni, Albert J. (2006) *Chemical and Biological Warfare: A Reference Handbook*, Santa Barbara, CA: ABC-CLIO

Heights of Ridicule: How the Hairpiece Brought Down a Kingdom

Blum, Stella (1982) *Eighteenth-Century French Fashions in Full Color*, London: Dover Books

Lacroix, Paul (1961) *France in the Eighteenth Century: Its Institutions, Customs and Costumes*, New York: Frederick Ungar Publishing

All in the Mind: The Quack Science of Phrenology

Tomlinson, Steven (2005) *Head Masters: Phrenology, Secular Education, and Nineteenth-Century Social Thought*, Tuscaloosa, AL, University of Alabama Press

Wyhe, John Van (2004) *Phrenology and the Origins of Victorian Scientific Naturalism*, Farnham: Ashgate Publishing

Hot and Bothered: Steam-Powered Cars

Setright, L. J. K. (2004) *Drive On!: A Social History of the Motor Car*, London: Granta Books

Sitting Pretty: The Crinoline

Lord, W. B. (2008) *The Corset and the Crinoline: An Illustrated History*, Mineola, NY: Dover Publication

Out of Breath: The Atmospheric Railway Experiment

Clayton, Howard (1966) *The Atmospheric Railways*, Lichfield: Howard Clayton

Hadfield, Charles (1967) *Atmospheric Railways*, Newton Abbot: David & Charles

Prindle, Steven (2006) *Brunel: The Man Who Built the World*, Phoenix Press

Gridlock: The Internal Combustion Engine

Lomberg, Micahel (2003) *Avoiding Gridlock*, North Mankato, MN: Smart Apple Media

Sikorsky, Bob (2006) *The Power of Green Driving: How We Can Get More Miles Per Gallon, Reduce Our Dependence on Imported Oil, and Curb Global Warming*, Tucson: Wheatmark

Humans and Superhumans: The Pseudo-Science of Eugenics

Agar, Nicholas (2004) *Liberal Eugenics: In Defence of Human Enhancement*, Malden, MA: Blackwell Publishing

Galton, Francis (1869) *Hereditary Genius – An Inquiry Into Its Laws And Consequences*, Galton Press

Kuhl, S. (2002) *The Nazi Connection: Eugenics, American Racism, and German National Socialism*, Oxford: Oxford University Press

Going Off With a Bang: Alfred Nobel and the Birth of High Explosives

Brown, G. I. (1998) *The Big Bang: A History of Explosives*, Stroud: Sutton Publishing

Bown, Stephen (2005) *A Most Damnable Invention: Dynamite, Nitrates, and the Making of the Modern World*, New York: Thomas Dunne Books

You Should Be So Lucky: Mint Cake, Cornflakes, and Other Happy Accidents

Simonis, D., ed. (2008) *Inventors and Inventions*, New York: Marshall Cavendish

Snake Oils: Radam's 'Microbe Killer' and Other Magic Potions

Hechtlinger, Adelaide (1974) *The Great Patent Medicine Era*, New York: Galahad Books

J.N. Hays (1998) *The Burdens of Disease*, Piscataway, NJ: Rutgers University Press

Palaces of the Skies: The Tragic Failure of the Zeppelin

Brooks, Peter (2003) *Zeppelin: Rigid Airships 1893–1940*, London: Putnam Aeronautical Books

de Syon, Guillaume (2007) *Zeppelin!: Germany and the Airship, 1900–1939*, Baltimore: John Hopkins University Press

Dick, Harold G. (1992) *Golden Age of the Great Passenger Airships: Graf Zeppelin and Hindenburg*, Washington, DC: Smithsonian

Toland, John (1972) *The Great Dirigibles*, New York: Dover Publications.

The Wonder Drug Gone Wrong: Heroin

Fernandez, Humberto (1998) *Heroin*, Center City, MI: Hazelden

Largo, Michael (2008) *Genius and Heroin: The Illustrated Catalogue of Creativity, Obsession, and Reckless Abandon Through the Ages*, New York: Harper Paperbacks

Plastic Not-So-Fantastic: The Depredations of Polythene

Fenichell, Stephen (1996) *Plastic: The Making of a Synthetic Century*, New York: HarperBusiness

Glowing in the Dark: The Deadly Craze for Radium

Clark, Claudia (1987) *Radium Girls: Women and Industrial Health Reform, 1910-1935*, Chapel Hill, NC: University of North Carolina Press

Curie, Eve and Sheean (1937) *Madame Curie: A Biography*, US: Da Capo Press

A Recipe for Disaster: Fast Food

Schlosser, Eric (2001) *Fast Food Nation: The Dark Side of the All-American Meal*, New York, NY: Houghton Mifflin

The First WMD: Tesla's Teleforce

Cheney, Margaret (2001) *Tesla: Man Out of Time*, New York: Simon & Schuster

Tesla, Nikola, (1937) *A Machine to End War*, www.pbs.org/tesla/res/res_art11.html

A Little Bit of What You Fancy: Trans Fat, HFCS and Food Additives

Shaw, Judith (2004) *Trans Fats: the Hidden Killer in our Food*, New York: Pocket Books

US Food and Drug Agency (2003) *Revealing Trans Fats*, www.fda.gov/FDAC/features/2003/503_fats.html

Colour Me Well: Spectro-Chrome Therapy

Wanjek, Christopher (2002) *Bad Medicine: Misconceptions and Misuses Revealed, from Distance Healing to Vitamin O*, Hoboken: Wiley

That Heavy Feeling: Leaded Petrol and Paint

Moss, Norman (2000) *Managing the Planet*, London: Earthscan Publications

Nevin, Rick (2000) *How Lead Exposure Relates to Temporal Changes in IQ, Violent Crime, and Unwed Pregnancy*, Environmental Research, 83

Chitty-Chitty Bang Crash: The Flying Car

Watts, Steven (2006) *The People's Tycoon: Henry Ford and the American Century*, London: Vintage

Monkeying Around: Quack Rejuvenation and Potency Therapies

Watkins, Elizabeth, W. (2007) *The Estrogen Elixir: A History of Hormone Replacement Therapy in America*, Baltimore, MD: The Johns Hopkins University Press

Flying High: The Discovery of LSD

Hoffmann, Albert (1979) *LSD: My Problem Child*, Santa Cruz, CA: MAPS

Lee, Martin A. and Shlain, Bruce (1994) *Acid Dreams: The Complete Social History of LSD: The CIA, the Sixties, and Beyond*, New York: Grove Press

Silent Spring: DDT

Carson, Rachel (1962) *Silent Spring*, New York: Houghton Miffin

Roberts, Donald (2009) *The Green Killing Fields: The Need for DDT to Defeat Malaria and Reemerging Diseases*, Washington, DC: AEI Press

Death Sprays: CFCs and Ozone Depletion

Sooros, Marvin (1997) *The Endangered Atmosphere: Preserving a Global Commons*, Columbia, SC: University of South Carolina Press

Walker, Jane (2004) *The Ozone Hole*, Mankato, MN: Stargazer Books

Barking Mad: Animals as Weapons of War

Cooper, Jilly (2002) *Animals in War*, Guilford, CT: Globe Pequot Press

Destroyer of Worlds: The A-Bomb

Cirincione, Joseph (2008) *Bomb Scare: The History and Future of Nuclear Weapons*, New York: Columbia University Press

Yep, Laurence (1996) *Hiroshima*, New York: Scholastic

Rhodes, Richard (1995) *The Making of the Atomic Bomb*, New York: Touchstone

Beyond Belief: L. Ron Hubbard's Dianetics

Barret, David B. (2007) *A Brief History of Secret Societies: An Unbiased History of Our Desire for Secret Knowledge*, New York: Carroll & Graff

Gallagher, E.V. (2006) *Introduction to New and Alternative Religions in America*, Westport, CT: Greenwood Press

Big Brother Is Watching You: Traffic Enforcement and Surveillance

Keenan, Kevin M. (2005) *Invasion of Privacy*, Santa Barbara, CA: ABC-Clio

Orwell, George (1949) *Nineteen Eighty-Four*, London: Penguin

Medicine or Poison: In Search of the Magic Bullet

Braithwaite, John (1984) *Corporate Crime in the Pharmaceutical Industry*, New York: Routledge

Meyers, Morton (2008) *Happy Accidents: Serendipity in Modern Medical Breakthroughs*, New York: Arcade Publishing

High Steppers: Platforms and Heels

Crane, Diana (2001) *Fashion and Its Social Agendas: Class, Gender, and Identity in Clothing*, Chicago: University of Chicago Press

Catch Me if You Can: The Birth of the Computer Virus

Okin, J. R. (2005) *The Internet Revolution: The Not-for-dummies Guide to the History, Technology, And Use of the Internet*, Winter Harbor, ME: Ironbound Press

Rosenberg, Richard (2004) *The Social Impact of Computers*, Bingley, West Yorkshire: Emerald Group Publishing

Format Wars: The Failure of Betamax

Haig, Matt (2005) *Brand Failures: The Truth about the 100 Biggest Branding Mistakes of All Time*, London: Kogan Page Publishers

Wasser, Frederick (2001) *Veni, Vidi, Video: The Hollywood Empire and the VCR*, Austin, TX: University of Texas Press

I Did it My Way: Karaoke

Xun, Zhou and Tarocco, Francesca (2007) *Karaoke: The Global Phenomenon*, Chicago, IL: Reaktion Books

Electronic Garbage: The Dissemination of Spam Email

Spammer-X, Posluns, Jeffrey (ed) (2004) *Inside the Spam Cartel: Trade Secrets from the Dark Side*, Rockland: Syngress Publishing

Electric Dreams: Human–Electric Hybrid Vehicles

http://www.sinclairc5.com (the site for Sinclair C5 enthusiasts worldwide)

Neither a Borrower nor a Lender Be: Subprime mortgages

Bitner, Richard (2008) *Confessions of a Subprime Lender: An Insider's Tale of Greed, Fraud, and Ignorance*, Hoboken, NJ: Wiley

Cooper, George (2008) *The Origin of Financial Crises: Central Banks, Credit Bubbles and the Efficient Market Fallacy*, Petersfield: Harmann House

Goodman, L., Li, S., J. Lucas, D., Zimmerman, T., Fabozzi, F. (2008) *Subprime Mortgage Credit Derivatives*, Hoboken, NJ: Wiley

Gramlich, Edward M. (2007) *Subprime Mortgages: America's Latest Boom and Bust*, Washington, DC: Urban Institute Press

Lowenstein, Roger (200) *When Genius Failed: The Rise and Fall of Long-Term Capital Management*, New York: Random House